T0258275

Industrial Policy,
Technology, and
International
Bargaining

ETEL SOLINGEN

Industrial Policy, Technology, and International Bargaining

DESIGNING NUCLEAR
INDUSTRIES IN
ARGENTINA AND
BRAZIL

STANFORD UNIVERSITY PRESS
STANFORD, CALIFORNIA

Stanford University Press
Stanford, California
© 1996 by the Board of Trustees of the
Leland Stanford Junior University
Printed in the United States of America

CIP data appear at the end of the book

Stanford University Press publications are
distributed exclusively by Stanford
University Press within the United States, Canada, Mexico, and
Central America; they are distributed
exclusively by Cambridge University Press
throughout the rest of the world.

To S., A., and G.

MEDIA GRAVITAS ORDONIS
TERRESTRIS MEI

Acknowledgments

Many more people than I am able to remember provided support during the exacting process of writing this book. The project as a whole would have hardly come to fruition without the consistent practical and intellectual sponsorship of the University of California's Institute on Global Conflict and Cooperation (IGCC). An IGCC Dissertation Fellowship facilitated my field research and the writing of the dissertation on which this book is based. For their advice and support as my doctoral committee, I am indebted to Edward Gonzalez, Richard C. Sklar, and Robert N. Burr. A grant from the University of California at Irvine allowed me, through a sabbatical leave during the fall of 1991, to begin transforming the dissertation into a book.

At different stages in the writing process, I have benefited from the comments and suggestions of David Becker, Norman Clark, Jim Dietz, Harry Eckstein, Albert Fishlow, Jeff Frieden, Bernie Grofman, Nora Hamilton, Michael Intriligator, Robert Kaufman, James Kurth, Sanford Lakoff, Heraldo Muñoz, David Pion-Berlin, William Potter, John Redick, David Rock, Michael Shafer, Kathryn Sikkink, Craig Smith, Kenneth Sokoloff, Dorothy Solinger, John Surrey, Scott Tollefson, Steven Topik, Norman Vig, Carlos Waisman, David O. Wilkinson, and Herbert York. I owe special appreciation to a long list of Brazilians and Argentines who, with their characteristic friendliness and frank opinions, were willing to share with me their own understanding of their country's nuclear program. These include Guido

Beck, Clovis Brigagão, Sergio Salvo Brito, Regis Cabral, Joaquim Carvalho, Jose Murilo de Carvalho, Marcelo Cavarozzi, Isidoro Cheresky, Renato Dagnino, Ubiratan D'Ambrosio, Walderley J. Dias, Jose Goldemberg, Daniel Heyman, Jorge Katz, Rogerio Cerqueira Leite, Witold Lepecki, Guillermo O'Donnell, Adilson de Oliveira, Oscar Oszlak, Luiz Pinguelli Rosa, Simon Schwartzman, Glaucio D. Ary Soares, Jorge Spitalnik, and Wilson Suzigan. From them, and others, I learned not only about differences between the two countries' politics and institutions but also about a common spirit. I am also indebted to governmental officials, engineers, technicians, and managers of private firms in both countries who provided me with an insight into their country's decision-making process and industrial strengths. I followed their request for anonymity, understandable in light of the programs' overall political sensitivity. In fact, this sensitivity made it enormously difficult to research the nuclear sector, even when industrial aspects (as opposed to strategic ones) were the focus of inquiry. Oftentimes I thought I might have been better off studying the toy industry. Without the stimulating and supportive infrastructure of the Instituto Universitário de Pesquisas do Rio de Janeiro (IUPERJ), and of Simon Schwartzman in particular, my field research would have been far less efficient. I am also grateful to Sara Tanis, Mario Mariscotti, and Amílcar Funes, from the Argentine Atomic Energy Commission, who shared with me a few retrospective thoughts.

The writing process would have been far less bearable without the support and affection of friends and colleagues at the University of California at Irvine. My deepest recognition for an incisive review goes to William Smith. I am also indebted to Carol Hartland and John Feneron of Stanford University Press for patient copyediting and to Senior Editor Muriel Bell for providing the most supportive and helpful editorial suggestions any author could hope for. Sections of Chapter 2 first appeared in *International Organization* 47, 2 (Spring 1993). Sections of Chapter 7 first appeared in Solingen (1990b).

Finally, my special appreciation to Fanny and the late Fito Shenhavi, above all, for being unique parents. I owe most to Simon, Aaron, and Gabrielle. Simon was always the best reference on anything having to do with technology; yet the book was—luckily—the most peripheral of our many joint undertakings, and my sole venture. My son, Aaron, and daughter, Gabrielle, offered me much-needed periodic refuge from industrial boilers and heat exchangers. They both magically renewed all the energy (nonnuclear) that the writing process drained. I dedicate this book to all three.

Notwithstanding the helpful contributions of all the individuals mentioned above, and the temptation to have them share at least some of the blame, the responsibility for any errors is mine alone.

E. S.

Contents

Figures and Tables

Preface

This book examines the sources of industrial policy and technological development. In this age of privatization and trade liberalization it is particularly appropriate to explore how interests— state and private, domestic and foreign—bargain over the gains and costs of industrial programs and why they arrive at different arrangements. Although the cases are drawn from the industrializing world, strategies aiming at increasing local content are widely embraced and are becoming central to foreign direct investment in the 1990s. The notions of promoting national private firms and supporting "strategic industries" to ensure a country's healthy technological development—the central theme of this book—have more than a familiar echo today. Such policy characterized Argentina's nuclear industry far more significantly than Brazil's, despite similarities in the two countries' overall developmental trajectory and endowments. This puzzling contrast carried over to differences in bargaining with foreign technology suppliers and in the alternative technological paths adopted.

In explaining these differences, I find it more useful to focus on domestic political structures and institutions rather than on market structures, international regimes, the political power of private entrepreneurs, or ideology. In Brazil, the regime's more or less consensual hierarchy of macropolitical goals and means, and the nuclear sector's segmented decision making, constrained the range of options: the nuclear industry was designed to follow the core parameters of Brazil's industrial model (late 1960s–early 1970s) and to help stem a fizzling economic miracle. This required—in the eyes of Brazil's decision makers—heavy reliance on state entrepreneurship and foreign technology, leaving less room for national private industrial and technological participation. The features of Argentina's nuclear industry can be understood, on the one hand, in the light of the country's characteristically low macropolitical consensus, and, on the other hand, in a high level of nuclear sectoral autonomy. Such autonomy stemmed from the tripartite division of the state among the armed services since 1955 and allowed the Navy-controlled National Atomic Energy Commission (Comisión Nacional de Energía Atómica—CNEA) to imprint the nuclear sector with its own institutional preferences for domestic private entrepreneurial and technical resources.

The argument I advance, building on the concepts of macropolitical consensus and lateral autonomy, explains policy over time, helps identify the conditions under which sectoral institutions matter, offers a guide to understanding how agencies formulate industrial options and preferences, and explains how the structures and institutions within which sectoral agencies operate affect bargaining processes with foreign actors. The wider relevance of macropolitical consensus and bureaucratic autonomy for industrial policy and outcomes is evident in examples drawn from a variety of industrialized and industrializing countries. Levels of consensus and institutional autonomy have consequences for international bargaining: they influence the size of domestic win-sets, the risks of involuntary defection, and the credibility of commitments. Macropolitical consensus and bureaucratic autonomy also affect entrepreneurial behavior, via the efficiency of political signals—about incentives and risks—that different combinations of consensus and autonomy breed.

I propose a fairly counterintuitive interpretation of the Brazilian and Argentine nuclear industries, considering the common tendency to explain such programs largely in geostrategic terms. Numerous articles and book chapters on these countries' nuclear programs have appeared in the general literature, but no professional book-length monograph in the English language has been devoted to a comprehensive analysis of this multifaceted sector. Most articles focus on the strategic implications of Brazil's and Argentina's pursuit of indigenous nuclear capabilities. Instead, this book studies nuclear programs in the context of industrial policy. This is not to say that strategic considerations are always irrelevant to the understanding of technical paths and industrial structures. Yet the development of an industrial capability in power reactor design and components does not invariably signal military objectives, although such capability lowers technical barriers when those objectives are embraced. In other words, a comprehensive industrial nuclear energy program is not *necessary* if military applications are the leading objective, because nuclear weapons can be obtained from a dedicated program on a much smaller scale and at less cost than those from a power plant industry. Neither is the existence of a large-scale industrial program *sufficient* to impute strategic intentions to the state that develops it: such intentions are the product of an express political decision to gear technical processes to military options.

Finally, the book questions the validity and limits of a relatively widespread contention (among developing countries) that nuclear technology "burns" industrial developmental stages through multiplier effects. This assumption, hitherto untested, is the subject of an entire chapter exploring the empirical, conceptual, and methodological difficulties embedded in the study of technological spin-offs. It is particularly worthy to note, in this context, that the nuclear programs of Argentina and Brazil accounted for over 5 percent of the two countries' respective foreign debt, among the highest in the developing world in the 1980s. The lessons from this experience may be highly valuable for their potential demonstration effect on other industrializing countries facing the task of bringing about technological modernization at rational social costs.

PART ONE

Explaining Industrial Strategy

CHAPTER I

Introduction

In defining their nuclear industries, Argentina emphasized the participation of its national private firms, whereas Brazil relied on state-owned enterprises and joint ventures with foreign firms. This contrasting division of labor is puzzling for three main reasons. First, both countries had a comparable history of state intervention and ownership.[1] Second, Argentina had fewer endowments (than Brazil) in the capital goods sectors relevant to nuclear power plant supplies. Third, the political strength of Argentine industrial entrepreneurs in those sectors was far weaker than that of their Brazilian counterparts. The latter were better organized, more cohesive, and politically more resourceful. The contrast in industrial structures appears counterintuitive because states are assumed to conform to the market (rather than displace private firms) more readily where those private firms are integrated and politically unified.[2] Conversely, the more fragmented, divided, and unorganized markets are, the more likely states are to expand into entrepreneurial functions. Yet Brazil displaced a more organized and cohesive private industrial sector through state-owned subsidiaries and joint ventures with foreign partners. Argentina, instead, promoted and protected national private firms in a relatively smaller, more fragmented, and less sophisticated industrial sector.

What makes this emphasis on domestic human and material resources even more intriguing is the overwhelmingly superior bargaining position of Brazil, as the emerging economic giant of the

1970s, in contrast to an unstable Argentine state with a declining industrial structure. Brazil's rising industrial and power capabilities would have allowed it to extract a larger role for its native firms had national endowments and state power mattered. Instead, the process of bargaining with foreign suppliers reflected a different set of choices about the structure of the nuclear industry. Those choices also involved different technological paths: light water reactors for Brazil, heavy water reactors for Argentina.

Beyond differences between the two countries in their emphasis on state entrepreneurship and in their bargaining with foreign suppliers, efforts by an industrializing state (in the 1960s and 1970s) to develop private entrepreneurship in the nuclear area seem, in themselves, an anomaly. Extensive state intervention and ownership was the norm throughout Latin America, particularly in energy markets and at the high-tech end of the sector.[3] In fact, state entrepreneurship in nuclear industries was common even in the industrial world because of the long lead time between technological choice and plant completion, the massive capital investments required, the long-term realization of returns, the high levels of commercial and technical risk, and the perception of nuclear energy's centrality to industrial growth.[4] From this perspective, an industrializing state's support for private ownership in the nuclear sector is a "least likely event"—and one suitable for a critical case study.[5]

The seemingly esoteric study of the development of a nuclear industry provides important clues about generic determinants of property relations between state and private (national and foreign) actors in different industrial sectors. Understanding the origins of these "mutual adjustments," or division of labor, is particularly useful now that privatization has become a reigning objective throughout the continent and beyond.[6] It is interesting to note, for instance, that some of the once small and often uncompetitive private partners in Argentina's nuclear program grew to become major beneficiaries of privatization in the 1990s. Earlier decisions about the industrial structure of a given sector helped transform the distribution of economic power and technical capabilities between the state and private sectors.

Brazil and Argentina provide us with an almost perfect set for a most similar systems design.[7] Both are latecomer, middle-income, industrializing capitalist states, with comparable levels of economic development, state intervention, techno-scientific competence, per capita gross domestic product (GDP), sectoral contributions to GDP, reliance on external sources of technology, capital, and equipment, and proneness to authoritarian rule through military intervention.[8] Furthermore, the international context provided both countries with similar political, financial, and commercial constraints and opportunities. How, then, do we make sense of these contrasting choices and bargaining behavior?

I find the answer to why certain sectoral arrangements emerge, and not others, rooted in variations in domestic structures and institutions. In particular, I argue that varying degrees of macropolitical consensus on the one hand, and of sectoral institutional autonomy on the other, can help explain differences in sectoral industrial policy.[9] Thus, high levels of macropolitical consensus in situations where sectoral institutions lack autonomy lead to sectoral policies that resemble broad industrial patterns. In Brazil, the regime's more or less consensual hierarchy of goals and means and a segmented decision-making process constrained the range of options in such a way that nuclear policy followed the core parameters of Brazil's industrial model: rapid growth through exports and macroeconomic stability.[10] State entrepreneurship and foreign technology had increasingly become the means, leaving less room for national private industrial and technological resources.

Instead, low levels of macropolitical consensus can help turn sectoral institutions with significant autonomy into the most critical research arena for understanding industrial policy. Such was the case in Argentina, where the tripartite division of the state among the armed services strengthened the autonomy of each in their industrial spheres of influence, allowing the institutional preferences of the Navy's National Atomic Energy Commission (CNEA) to prevail. These preferences led to the Commission's special emphasis on domestic private entrepreneurial and technical resources.

Explaining Industrial Choices

Why and when do states nurture and protect national private firms? Why and when do states displace them, either by stepping in directly as producers or by purchasing equipment from foreign competitors? Different conceptual traditions provide competing interpretations of this variability in state intervention. A prime suspect is always *market failure*, that is, states intervene where private firms are reluctant to invest. Yet market failure may explain generic state intervention in the establishment of a nuclear industry as a whole, but not differences in the extent, nature, or instruments of state entrepreneurship.[11] Moreover, the private sector in both countries was equally willing to supply inputs for nuclear reactors and fuel-cycle facilities, given the right inducements. Yet public ownership was more pervasive in Brazil.

A second hypothesis involves a simple *microeconomic* explanation, according to which the country with the more efficient set of private firms displays higher shares of participation by those firms. Conversely, the less competitive the private sector, the higher the likelihood that state or foreign firms will replace it. These expectations are overturned by the reality of more competitive Brazilian firms—relative to their Argentine counterparts—seeking a share of the inputs for nuclear reactors.[12] An independent probing by the Bechtel Corporation in 1973 confirmed the existence of a solid Brazilian infrastructure in electromechanical and engineering services and equipment. The private firms that composed this infrastructure were estimated to enable a "national content" level of close to 70 percent for the first nuclear power plant, to be built through a large-scale program beginning in 1974. Argentine firms, in contrast, were far less equipped, particularly in the mid-1960s, when the feasibility studies for the country's first nuclear reactor were conducted. As a result, and in spite of a dedicated effort by CNEA to engage private firms, these firms were able to contribute only about 13 percent of the total electromechanical supplies. Clearly, the relative competence and efficiency of the private sector in each country cannot account for the outcome. Efficiency would have predicted less state entrepreneurship and foreign supplies where national private firms were better endowed, as in Brazil. Instead Brazil created public enterprises, whereas the Argentine state played a more subsidiary role.

A third hypothesis, an extension of the previous one, invokes *international market considerations* in explaining state intervention and the resulting adjustment among state, foreign, and local private firms.[13] Distributional outcomes among these three sectors can be traced to the degree of vulnerability of national firms to international competitors. Where domestic firms are less competitive, they are more likely to be displaced by foreign ones. Following this reasoning, we would expect Argentina, with the most vulnerable private sector, to have been saddled with a higher ratio of foreign supplies. The outcome, however, challenges that expectation, as is clear from the detailed comparison of respective levels of domestic content in Chapter 2.

The external vulnerability of private local firms is also hypothesized to lead to the creation of public enterprise, either because the state cannot protect domestic producers or because it views public enterprise as a means to compete internationally, or both.[14] This logic should have led to more state entrepreneurship in Argentina than in Brazil, whereas actually state subsidiarity (*subsidiaridad del Estado*) was more characteristic of Argentina.[15] This outcome questions the link between state ownership and external vulnerability. In fact, both countries aimed at becoming major emerging exporters of nuclear technology and equipment, yet Brazil relied primarily on newly created state-foreign joint ventures, whereas Argentina promoted its private firms.[16]

Paradoxically, international market considerations might have acted—in both cases—to broaden the opportunities for a higher share of inputs from domestic firms, both public and private. The review of the global market for nuclear reactor technology and components found in Chapter 4 clearly challenges the proposition that international market forces are invariably able to impose their predatory interests on vulnerable firms in developing countries. A growing competition among suppliers, excess capacity, and the high financial stakes of each transaction in the shrinking market of the 1970s enabled recipients, many of them newly industrializing countries (NICs), to extract a maximum level of participation for their private firms as a condition for granting contracts to foreign nuclear architect-engineering firms. Moreover, abundant foreign financing during those years offered greater leverage for recipients to resist supplier credits; such credits often displaced local firms by tying financing to the purchase of equipment. Yet Brazil—a favorite of international finance—was less prone than Argentina to tie contracts to the participation of its private firms. Clearly, international market considerations are underdetermining, since they cannot explain why Argentina exploited such permissive conditions while Brazil did not.

A fourth hypothesis might be marshaled to explain the contrasting responses of Brazil and Argentina, by focusing on how the *political strength* of national private firms may have affected their share of supplies. According to this classic "interest group" explanation, the more politically robust the specific sector, the higher the probability that it will be protected from the expansion of state firms or foreign competitors. Robustness here is a function of how concentrated and organized, as opposed to how diverging and divided, the interests of the private firms are. As Samuels (1987) suggests, states are assumed to more easily displace unorganized private firms than a highly integrated sector. Yet Nuclebrás, the state holding company, expanded in Brazil at the expense of a quite cohesive group of capital goods and engineering firms. Capital goods producers in Brazil were led by a small number of very large—tightly organized—firms, highly dependent on state orders.[17] Among the potential suppliers of the nuclear program were Brazil's largest industrial groups, such as Villares and Bardella. As major supporters of the military regime since its inception in 1964, the sector's political strength increased in tandem with its economic expansion, particularly since the early 1970s.[18] The Brazilian Association for the Development of Basic Industries (*Associação Brasileira para o Desenvolvimento das indústrias de base*) (ABDIB), which represented many of the firms with potential involvement in the nuclear program, was particularly strong. In Argentina, however, the Atomic Energy Commission faced a much less unified and politically powerful group of firms. Those with potential links to the nuclear program were not very influential in the 1960s within the dominant, liberal-minded sectoral organization Unión Industrial Argentina (UIA).[19] The relative political strength of industrial firms, therefore, fails to explain why powerful Brazilian entrepreneurs received contracts for less than 50 percent of what they were able to supply, while their feebler, mostly nascent Argentine counterparts were allocated the maximum possible participation. Chapter 3 explains the foundations of the relationship between the state and private economic actors in each case.

A fifth hypothesis for differences in the division of labor between national (state and private) and foreign firms involves *international power* considerations. From this vantage point, an internationally more powerful state is arguably in better condition to protect its domestic interests. Structural power is often defined as an aggregate of political, military, and economic components of power that determine a state's ability to influence others' policies.[20] Domestic sectoral endowments and characteristics are marginal to this argument, whereas overall state position in the global system is primordial. Yet, had state power been relevant, one would have expected Brazil to be better able to protect its private firms, by virtue of these firms' greater industrial strength and of the country's overall stronger bargaining position. The strength of Brazil's bargaining position can be deduced from three facts. First, the negotiations involved eight nuclear plants and related fuel-cycle facilities. "No agreement" threatened the survival of the supplier Kraftwerk Union because of the small, zero-sum market of the 1970s. Second, Brazil was West Germany's most important economic partner in the developing world. Lastly, Brazil had greater internal stability than Argentina, thus providing a safer environment for foreign investments.

Summing up, Brazil was a rising economic power in the 1960s and 1970s whereas Argentina's dramatic decline became the riddle of development theories.[21] Yet the weaker of the two states was able to secure a relatively better position for its private firms and to exact greater concessions from nuclear suppliers. Structural power capabilities thus fail in this case to predict overall levels of domestic participation and have little to say about resulting mixes of state and public entrepreneurship. The generic deficiency of mercantilistic or neorealist arguments is that states' relative power is not always a good guide for identifying state interests. Understanding the origin of the different choices provides a far better approximation for assessing and explaining bargaining outcomes.

Finally, *ideological* considerations are at times advanced to explain different industrial choices despite similar international constraints and opportunities.[22] Argentina's choices, for instance, are often traced to a nationalist ideology geared to develop an autonomous nuclear capacity. Yet both Brazilian and Argentine "nucleocrats" shared such ideology fundamentally, while opting for different technical and industrial paths. Self-reliant nationalism was never the monopoly of the Argentine Atomic Energy Commission; it was deeply rooted among technocratic groups in other sectors, such as pharmaceuticals and aerospace. It was alive in Brazil's Nuclear Energy Commission as well. However, it did not lead to similar outcomes—arrangements among foreign, national private, and state firms—across these sectors. Ideology in itself, therefore, can hardly explain differences between the Brazilian and Argentine nuclear industries, or among Argentina's different industrial sectors. As an explanation, ideology is clearly not sufficient and, at times, not even necessary.[23] Its fundamental weakness is its inability to predict outcomes by itself; at best, it is always subsidiary to other explanations.[24]

The explanation I develop in Chapter 2 focuses on variations in domestic structures and institutions. On the one hand, identifying the relevant structures or policy networks is a necessary first cut.[25] As argued, "the choice of public enterprise . . . is the outgrowth . . . of a particular constellation of power relationships."[26] On the other hand, the institutional setting mediates the influence of social forces or defines the framework within which politics takes place.[27] An analysis of institutions and procedures helps single out the locus and characteristics of the decision-making arena and its guiding preferences.[28] What structures of opportunity may a sectoral institution seize in defining industrial policy? Why and how does such an institution become a central research arena, and what difference does it make?

Macropolitical Consensus, Bureaucratic Autonomy, and Industrial Policy

I propose that varying degrees of macropolitical consensus on the one hand, and of autonomy of the sectoral agency on the other, can explain industrial policy.[29] Macropolitical goals are those at the apex of the state's hierarchy of goals and means.[30] They need to be defined more explicitly than simply as the pursuit of economic growth or national security. General means are often embedded in the definition of macropolitical goals. Thus, they may involve the pursuit of export-led growth, or of macroeconomic stability, or of a more egalitarian income distribution, or the strengthening of national private capital. Macropolitical goals often reveal different definitions of property rights and of their distribution (between state and private agents and between local and foreign agents, for instance). Such goals can be deduced from an analysis of policy options most likely to lengthen the regime's longevity. They can also be identified through a review of declaratory policy and informal statements by the ruling coalition and its supportive networks. There are methodological trade-offs between these two independent measures; where they converge, our certainty about the nature of macropolitical objectives grows.

Macropolitical consensus is the expression of widely shared preferences over macropolitical goals among major political actors.[31] Such consensus grows when a dominant coalition with homogeneous, converging interests monopolizes political power. The consensus may include private actors even when they are not central to the policy process.[32] Levels of consensus can be assessed by identifying the most critical components of the coalition—whose core macropolitical goals converge—without which no such coalition is possible. A stronger macropolitical consensus is more likely where all or most critical components—those with veto power—are "inside the tent." This is the equivalent of a "historic bloc," which, by including all critical components, may be better able to formulate a more cohesive, stable, raison d'état.[33] It is possible but not inevitable that the sequential inclusion of additional coalition members may add up to a weakening consensus, particularly if the homogeneity regarding objectives is diluted; the larger the numbers, the higher the probability that new objectives and demands will be added. In other words, there may be a trade-off between the extensiveness and the intensity of the consensus.[34] At times societal sectors—labor, for instance—are excluded to strengthen consensus. Finally, the longevity of a coalition—the length of the expressed a priori willingness of members to support its policies—can be a measure of consensus. Very often, continuous shifts in membership in the coalition, resulting cabinet changes, and an unstable course are symptoms of a very fragile consensus based on a "moving equilibrium" of highly competitive political and economic groups.[35] Longevity of commitment, however, is not always easy to identify and is neither a necessary nor a sufficient condition for a strong consensus to exist. Above all, consensus is always a matter of degree; it rarely implies complete harmony of interests and it can be best identified relative to other states or historical situations.[36] While "conflictual outbursts in the midst of consensus" characterized Japanese politics in the postwar era, brief episodes of consensus sprinkled a highly conflictual political history in Argentina.[37]

Examples of high macropolitical consensus, of varying contents, include Germany and Japan, particularly after the immediate postwar era. The critical components of Japan's grand coalition were the industrial *keiretsu* barons that financed the Liberal Democratic Party, and the state bureaucracy. The content of that consensus was high rates of economic growth through the promotion of exports, state intervention, sectoral policy, and the regulation of imports. The South Asian NICs revealed a similar basis of consensus, until very recently. There was also high consensus—albeit of a different nature—in India and Israel following independence and until the late 1970s, and in Chile in the 1970s and 1980s. The politics of smaller

European states, such as Norway and Sweden after the 1930s, or Austria and Switzerland, have also been characterized as highly consensual.[38] The Czech Republic's evolution from a command economy to a market economy appears to be steered by a highly consensual political transition.

Particularly low levels of consensus have characterized postwar Argentina, Germany's Weimar Republic, and, perhaps, the United States at the end of the Reagan era. The much-discussed "decline of American hegemony" fueled a debate over the need, and the appropriate formula, for an industrial strategy.[39] Israeli politics from the late 1970s to the early 1990s has also shown the strains of low consensus over industrial policy, as well as over territorial compromise in the West Bank and Gaza. The postindependence difficulties in the Ukraine were largely a symptom of extremely low consensus over the preferred path from a highly centralized economy into a market-oriented one.

What are the sources of macropolitical consensus? Why is it prevalent in certain situations and not others? In East Asian countries, external security threats, a relatively egalitarian income distribution, and a benign form of U.S. hegemony are often invoked to explain consensus over industrial policy.[40] The first two explanations are also applicable to the post-independence consensus in India and Israel. In general terms, high levels of state autonomy are a key ingredient in strengthening consensus over macropolitical goals across state institutions.[41] Rueschemeyer and Evans (1985), for instance, argue that effective state intervention may grow initially out of coherent bureaucracies relatively autonomous from dominant social interests.[42] Similarly, Fishlow (1987) establishes a connection between national security, state autonomy, and the coherent industrial project of East Asian countries. In many cases the exclusion of popular sectors is assumed to contribute to such consensus by narrowing the dominant coalition.[43] Economic crises, such as prolonged stagnation, are another potential source of consensus.[44] In particular, such crises tend to create coalitions that persist over longer periods of time.[45] Finally, international institutions can impose consensus as a condition for delivering relief or aid. The International Monetary Fund (IMF) and World Bank have acted in many countries to catalyze consensus around economic orthodoxy and structural adjustments. Thus, in different contexts, and occasionally in coexistence, external factors (security threats, a benign form of hegemony-cum-protection, international institutions) and internal ones (high levels of state autonomy, economic crises, the prospects of a relatively egalitarian income distribution) strengthen consensus. The examples of high and low consensus discussed earlier point to these sources of consensus.

Whatever its source, the degree of macropolitical consensus has important consequences for the operation of state bureaucracies. On the one hand, where macropolitical goals are contradictory, interagency bargaining increases and bureaucracies become balkanized.[46] Low consensus may lead to balkanization by straining the internal hegemony of core institutions—such as the Finance Ministry—within the state apparatus. This hegemony acts as a transmission belt by projecting the content of consensus onto—and throughout—state agencies.[47] The strategic position of the British Treasury within the bureaucracy in the 1930s is an example of that hegemony, as are MITI in Japan and the French Commissariat du Plan during the early postwar era.[48] In each case, a decline in macropolitical consensus led to decreasing central influence and greater balkanization. In other words, low consensus often reduces the ability of central agencies to influence the parameters of sectoral decision making. Consequently, in light of weakened central guidelines, the latitude of sectoral institutions to pursue a wider range of policy options grows to include all logical possibilities—technical and economic—regarding foreign and domestic inputs, public and private procurement, subcontracting, research and development (R&D) activities, and financing arrangements. In particular, such agencies are more likely to seek societal clients and to pursue particularistic interests.[49] In sum, ambiguity in macropolitical goals (low consensus) expands the range of permissible options in sectoral industrial policy.

On the other hand, where there are high levels of macropolitical consensus, such consensus is often expressed through a bureaucratic machinery with a more or less homogeneous outlook—an "embedded orientation" in Bennett and Sharpe's (1985) terms—over industrial policy. Bureaucratic guidelines thus tend to be more coherent, as in the early postwar years in Japan and among East Asian NICs, and the range of options is more limited.[50] This "trimming of the edges" in the repertoire of options may preclude an agency from embracing a policy it might have otherwise preferred, following its own "local rationality."[51] If a nuclear agency's "local rationality," for instance, compels it to strengthen a domestic constituency likely to support the program's continuity and expansion, it may pursue an indigenous technical capability and resist foreign equipment. Agencies will be more likely to bend their preferences and yield to central priorities when the priorities are highly consensual, because the agencies' authority can be challenged more effectively under such conditions.[52] Coherence in bureaucratic guidelines does not imply absence of goal conflict, which is inherent to any organization, including the state. Yet the presence of more or less consensual macropolitical goals may impose a series of overarching choice constraints on state agencies. A commitment to export-led growth may be just such a constraint and could preclude the inclusion of national private firms when the inclusion weakens such commitment. We can thus summarize the consequences of degrees of macropolitical consensus for the operation of state agencies by postulating that high consensus tends to constrain, whereas low consensus enables.

The ability of a sectoral agency—such as Nuclebrás or the Argentine Atomic Energy Commission—to take advantage of a broadened repertoire of options depends on its degree of autonomy, both vertical and lateral. Bureaucratic autonomy has generally been defined as "relative undisputed jurisdiction" over a function or a service.[53] At the high end of the spectrum, such agencies are endowed with quasi-sovereign powers.[54] An agency with high levels of *lateral autonomy* has greater capacity to define and carry out policy independently of the interference of other units. The more an agency can monopolize most aspects of a program—including planning, financing, technology transfer negotiations, training, fiscal incentives, and safety regulations—the higher its lateral autonomy. That is, autonomy grows as other agencies are effectively precluded—by formal procedure or by political arrangements—from influencing the choice of policy. Lateral autonomy can shield a unit from bureaucratic crosscutting pressures from other state agencies; it is thus a relational property of a state unit with respect to others and should not be confused with state autonomy as a whole.[55]

The logical opposite of lateral autonomy is *lateral segmentation*, where a number of agencies, ministries, and state firms have overlapping jurisdictions over either the definition of a certain policy or its implementation.[56] The greater the number of agencies involved, the higher the segmentation.

Examples of agencies with lateral autonomy include the U.S. Federal Reserve Board and the German Bundesbank, and many atomic energy commissions throughout the world, particularly in their early years. Encarnation and Wells (1985) identify laterally autonomous agencies in the developing world, in oil exploration, petrochemicals, export-processing zones, and computers. Levels of lateral autonomy, of course, can vary over time. The origins of lateral autonomy may have something to do with how and whether autonomy is maintained. Budgetary independence can strengthen autonomy, as is the case with some U.S. agencies that rely on user fees or earmarked taxes, or with programs over which Congress has less discretion over levels of funding. The Federal Reserve has an entirely independent source of funding, not subject to the standard Congressional appropriation process. Budgetary independence can be achieved outside of formal procedures, as when the U.S. National Security Council under the guidance of Lt. Col. Oliver North sought external funding for the agency's autonomous operations. National security has been a well-known source of lateral autonomy during the Cold War era in many countries. Institutional growth in size and complexity is also a source of autonomy.[57] A monopoly over the tasks involved, particularly when an agency controls downstream policy implementation, may increase autonomy as well.[58]

More generally, the lateral autonomy of certain agencies can be traced to historical-political compromises and is not always formally enshrined. For instance, certain Israeli ministries and agencies have enjoyed significant lateral autonomy as a result of the country's electoral system. Proportional representation, and the fact that no single party in 44 years has been able to get the required majority to govern on its own, has left small fringe (religious) parties with the power to extort such autonomy for the agencies they control. Under such conditions, attempts by the legislature (Knesset) to obtain proper information regarding budgetary allocations and procedures relative to those agencies rarely bore fruit. Other, quite peculiar, historical compromises explain the lateral autonomy of military industries and arms exports in China.[59] In Argentina, three highly antagonistic armed services partitioned the state and its associated industrial sectors among themselves in 1955. In that process, economic institutions controlled by the Army, the Air Force, and the Navy gained high levels of lateral autonomy. As many of these examples suggest, where lateral autonomy is granted at the moment of institutional creation, agencies are more likely to maintain it. The phenomenon of lateral autonomy is thus of wider incidence than one might suspect, and the conditions under which it makes a difference need to be specified.

The degree of an agency's *vertical autonomy* can be measured by

the extent of its formal or informal accountability to the top executive. Where the executive has limited jurisdiction over the agency's
preferences, or is expected to rubber-stamp them, the agency's vertical autonomy grows. Rubber-stamping is often a function of the technical complexity of the agency's portfolio or of a practical political
compromise, which precludes the executive from seriously challenging the agency's prerogatives. These limits on the agency's vertical
accountability operate in many of the cases of lateral autonomy discussed previously. Some of the conditions leading to lateral autonomy tend to tame effective control by the top executive as well.[60]
The logical opposite of an agency's vertical autonomy is *vertical centralization*, where decision making is effectively concentrated at the
top executive level.[61] In sum, a highly autonomous agency enjoys
horizontal control over policy and its implementation and is largely
unencumbered by a superior authority. Such was the case with the
Argentine nuclear agency (CNEA), an institution unhampered by the
agendas of energy or economic bureaus, and only formally accountable to the nation's president. In practice, CNEA had ultimate power
over most decisions and over implementation.[62] In contrast, Brazil's
nuclear decisions were made and implemented by the top executive,
in a vertically centralized but horizontally segmented context, where
the nuclear agency had little autonomy.

Low macropolitical consensus and a high incidence of lateral institutional autonomy among state agencies are mutually reinforcing.
However, low consensus does not always imply—although it often
enables—autonomous units; such units (central banks, for instance)
may also coexist in arenas characterized by highly consensual
macropolitical goals. Similarly, when high consensus permeates
state agencies, it is easier to relax vertical centralization. Yet agencies can maintain their vertical autonomy under low macropolitical
consensus as well. As this discussion suggests, macropolitical consensus is neither a necessary nor a sufficient condition for lateral or
vertical autonomy. Such autonomy can be rooted in enduring institutional characteristics of the state, which are not easily assailed by
cycles of higher or lower macropolitical consensus. Institutions, as
March and Olsen (1989:159) argue, are not simple echoes of social
forces.

What are the implications of an agency's lateral autonomy for
policy outcomes? Lateral autonomy neutralizes classical bureaucratic politics and allows the agency's idiosyncratic interests—such

as a preference for market solutions or state subsidiarity—to flourish.[63] The greater the agency's budgetary (material), functional (technical, information), and hierarchical (power-related) independence, the greater its opportunity to define a sector according to its "local rationale."[64] Segmentation, in contrast, imposes checks and balances and prevents the dominance of particularistic orientations; in other words, it thwarts preferences that may deviate from the accepted boundaries of industrial policy. High segmentation can be to industrial policy what polycratic chaos—rivaling ministries and incoherent allocation of resources—can be to the formulation of grand strategy.[65] Yet the presence of lateral autonomy has no intrinsic implications regarding either the efficiency or the desirability of a given policy. The high interest rates imposed by the autonomous Bundesbank on the rest of Europe may be defended as sound anti-inflationary policy from one perspective, and as German pursuit of egoistic self-interest from another. Lateral autonomy allowed President Reagan's National Security Council to pursue policies that contradicted legal procedures and to transgress the constitutionally accepted boundaries of action over Central American policy at the time. Such autonomy can, in fact, lead to rogue sectoral programs as often as it ensures relatively effective decision making, free of external bureaucratic constraints. Segmentation, in turn, may contribute to a more comprehensive consideration of most alternatives and consequences.[66]

We are now in a position to assess the research implications of different mixes of macropolitical consensus and bureaucratic autonomy (Figure 1). The impact of vertical autonomy on cell characteristics is more marginal. Beginning with the lower right-hand cell (IV), to the extent that consensus is low and sectoral agencies lack lateral autonomy, explanations based on bureaucratic politics—pulling and hauling among various agencies—may be particularly useful. The outcome in this situation is often an incoherent policy because it is formulated and implemented in the context of unstable or cyclical central priorities on the one hand, and of clashing bureaucratic institutional preferences on the other. The existence of powerful, antagonistic, private clienteles lobbying different agencies often exacerbates inconsistency and immobilism. Levels of vertical autonomy rarely alter the fundamentally contested nature of decision making in such cases, given low macropolitical consensus and high institutional segmentation.[67]

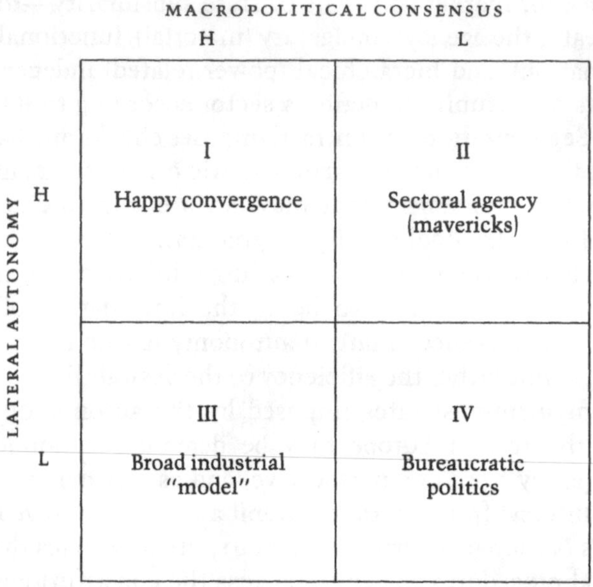

MACROPOLITICAL CONSENSUS

	H	L
H	**I** Happy convergence	**II** Sectoral agency (mavericks)
L	**III** Broad industrial "model"	**IV** Bureaucratic politics

LATERAL AUTONOMY

FIG. 1. The implications of macropolitical consensus and lateral autonomy for explaining industrial policy. Source: Solingen (1993a:283).

Where macropolitical consensus is relatively high and lateral autonomy low (cell III), sectoral policy will more closely resemble broader patterns (or the state's generic industrial model) at whatever levels of vertical autonomy. In this case, industrial-technological goals at variance with such consensus are not likely to survive.[68] Efforts at untangling sectoral decision making should be, therefore, directed at understanding the makeup of the dominant coalition and its core preferences. This will also be the case where there are high levels of both consensus and autonomy (cell I), provided there is a happy convergence between broad industrial patterns and the agency's preferences.[69] It may be much harder to predict the nature of sectoral policy in cases where an autonomous agency's preferences diverge from the policy set covered by the consensus.

Finally, when macropolitical consensus is low and lateral institutional autonomy high (cell II), an analysis of the sectoral agencies' institutional objectives, interests, trajectories, and ideology may provide a useful shortcut to our understanding of policy choice. This is often the case with a maverick agency shaping an industrial sector almost single-handedly, particularly when its vertical autonomy is also high. The counterintuitive consequence (at least from a bureau-cratic politics perspective) of such circumstances is that the most powerful state institution—a Ministry of the Economy or a MITI, for example—does not necessarily prevail in policy-making. The Atomic Energy Commission was not the most powerful institution within the Argentine state; it coexisted with other powerful agencies but enjoyed lateral and vertical autonomy. Low bureaucratic autonomy can tame a sectoral institution's idiosyncratic character-istics, *if* these depart from those of its bureaucratic sovereign.

Two observations are in order. First, although the degrees of macropolitical consensus and lateral autonomy—in themselves—may not predict policy outcomes, they can foreshadow the explana-tory strength of a sectoral agency's institutional interests and pecu-liarities, or of bureaucratic politics, or of the overall industrial model. In other words, knowledge about levels of consensus and autonomy helps us specify the locus of "external brokerage," or mediation among local firms and foreign capital, markets, and technology.[70] The specific outcome will be a function of the content of the consen-sus and of the agencies' preferences (where these are relevant). Sec-ond, knowledge about institutional preferences may help explain the content of certain choices but not why such choices prevailed. Hence, both the structural and institutional contexts in which the agency operates must be internal to the explanation of industrial policy.

What does all this tell us about nuclear choices in Brazil and Ar-gentina? In Brazil, a more or less consensual hierarchy of goals and means and a segmented decision-making process (cell III in Figure 1) constrained the range of options in such a way that nuclear policy followed the core parameters of Brazil's industrial model: rapid (par-ticularly export-oriented) growth and macroeconomic stability. State entrepreneurship and foreign technology had increasingly become the means toward that end, leaving less room for national private in-dustrial and technological resources. In Argentina, a low macropol-itical consensus and the tripartite division of the state among the armed services strengthened the autonomy of each service within its industrial sphere of influence. This situation broadened the services' range of options, allowing the institutional preferences of the Navy-controlled Atomic Energy Commission to prevail (cell II). These pre-ferences led to the Commission's special emphasis on domestic pri-vate entrepreneurial and technical resources. A group of private firms selected to provide equipment and services for the nuclear program, and a relatively large scientific-technical community, thus became the Commission's major political constituencies. This industrial blueprint did not merely buttress the Commission's position within the delicate balance of power among the services, but was also compatible with the Navy's classical support for business, state subsidiarity, and technical achievement. These differences between Brazil and Argentina shaped their respective win-sets in the bargaining game with foreign suppliers and were simultaneously affected by that game.[71] Chapter 4 examines this interplay between domestic structures and institutions and the external bargaining process.

Scope, Objectives, and Findings

This book has two main objectives. The first—explaining the sources of policy differences between countries and across sectors— is the subject of Parts I and II. The argument I advance in these more theory-laden parts of the book has implications for three important debates in international political economy. The first debate involves the nature of the relationship between states and private economic actors and how that relationship influences industrial policy and technological choice. The second debate concerns the relative importance of external (international political, market, and financial) factors, as opposed to internal (domestic structures and institutions) influences on decision making, bargaining, and outcomes. In particular, the account I propose here regarding the sources of industrial policy—macropolitical consensus and bureaucratic autonomy— sheds light on the process of bargaining between technology recipients and multinational corporations. The argument thus helps untangle the conditions under which recipients may exploit the opportunities provided by the international market, even in high technology, and despite the inherently predatory nature of such markets. Thirdly, in the context of the debate over interactions between technology and politics, I adopt a perspective of technological change as a dependent variable, a product of particular configurations of interests and values.[72] In exploring how choices of industrial structure may define technological paths, my argument reverses a more traditional concentration, in studies of this genre, on the impact—rather than the sources—of technological decisions.[73]

The second objective is to evaluate the Brazilian and Argentine nuclear industries from the perspective of technological change among NICs.[74] Technology is at the heart of industrial development and has been regarded as a major bottleneck in the economic evolution of industrializing countries.[75] The main question I address in Part III is the extent to which differences in initial choices regarding industrial structure and technological paths shaped the development and industrial-technological impact of nuclear programs in each country. In other words, Part III complements the analysis by specifying the relationship between policy choice and outcomes. The findings reported in Part III are relevant to the general debate over the effects of foreign technology on the development of technological capabilities among latecomers. They also shed light on specific claims about spin-offs from nuclear industries, often voiced by NICs. The chapters in this section follow the process of technological diffusion at all levels of analysis: international (from multinationals to recipients), state (technology screening, regulatory and integrative functions), and firm (technological learning).

I examine the impact of macropolitical consensus and lateral autonomy throughout five different historical conjunctures in Brazil and Argentina, thus broadening the number of observations where the argument can be tested.[76] Moreover, the final chapter explores the generalizability of the main theoretical claims by applying them to a broader range of industries, in the energy sector and beyond, in both countries.

The study yields the following general propositions.

On the sources of nuclear industrial policy:

1. Neither the level of private sector endowments, nor the political strength of industrial entrepreneurs, accounts for the different initial choices of nuclear sector structures. Changes in macropolitical consensus and lateral autonomy increased the influence of these two factors subsequently.

2. Knowledge about levels of macropolitical consensus and sectoral autonomy helps anticipate the explanatory relevance of broad industrial models, sectoral institutions, clientelistic networks, or bureaucratic politics, as critical research arenas.

3. The degree of macropolitical consensus has important consequences for the operation of state bureaucracies and for industrial policy: high macropolitical consensus constrains the range of sectoral choices, whereas low consensus enables sectoral institutions to pursue their own local rationality.

4. Where both consensus and sectoral institutional autonomy are low, explanations based on bureaucratic politics are particularly useful.

5. Where consensus is high and autonomy low, sectoral policy is more likely to resemble the generic industrial model.

6. This will also be the case where both consensus and autonomy are high, provided there is a happy convergence between the generic industrial model and sectoral institutional preferences.

7. Where consensus is low and autonomy high, an analysis of the sectoral agencies' institutional interests and trajectory will be most useful.

8. Levels of macropolitical consensus and sectoral autonomy affect the behavior of private entrepreneurs by presenting them with disparate sets of incentives and risks. The efficiency of political signals is highest where bureaucratic autonomy is high, even if macropolitical consensus is low. Signals are also efficient if consensus is also high, provided a happy convergence reigns between central agencies (broad industrial patterns) and sectoral ones. Low bureaucratic autonomy and low macropolitical consensus provide the context for least efficient signaling. Low autonomy and high consensus may allow more efficient signaling than in the previous case, although a high number of agencies may inhibit optimal efficiency.

9. Levels of macropolitical consensus and bureaucratic autonomy explain why different patterns of state-scientists relations emerge across industrial and scientific sectors.

On the implications for bargaining with technology suppliers:

10. Bargaining advantages cannot be simply deduced from international structural conditions. Identifying the respective win-sets requires prior knowledge of the levels and content of macropolitical consensus and bureaucratic autonomy, both of which shape the bargaining context in a very fundamental way.

11. Levels of bureaucratic autonomy and macropolitical consensus can influence the process of bargaining with technology suppliers through their impact on the size of domestic win-sets, on the risks of involuntary defection, and on the credibility of commitments and reduction of uncertainty.

12. Low consensus and high levels of autonomy narrow the size of the win-set to the institutional preferences of the sectoral agency. A small domestic win-set, in turn, can be a bargaining advantage and can increase the negotiators' leverage over the distribution of bene-

fits from the international bargain. High levels of lateral autonomy, even in the midst of low consensus, cancel out the risk of involuntary defection, the small size of the win-set notwithstanding. Thus, a state with lower levels of consensus over industrial policy, and a politically "weak" one at that, may be able to extract greater concessions from foreign suppliers than one with a more coherent industrial strategy.

13. Bargaining advantages dissipate when both consensus and autonomy are low, because this combination increases uncertainty regarding the contours of the win-set, raises the risk of involuntary defection, and weakens the credibility of commitment. These conditions thus breed unstable demands, forcing the technology supplier to reassess continuously the interplay among political and bureaucratic forces within the recipient state. The bargaining process becomes protracted and unpredictable, leading suppliers to require additional assurances (side-payments) that ratification will take place.

14. The independent effects of high macropolitical consensus are mixed. On the one hand, high consensus in actor A may dissipate fears of involuntary defection in the opponent B, because there is greater certainty that A "can deliver," (which, in turn, strengthens A's bargaining position). On the other hand, because high consensus improves the chances of easy ratification, A's negotiators are less able to use domestic pressures as a bargaining chip to obtain growing concessions from B. The effects of high consensus, therefore, may be better gauged in conjunction with levels of bureaucratic autonomy.

15. Conditions of high consensus and high bureaucratic autonomy increase the risks of involuntary defection, make the environment less predictable, and weaken the advantages of consensus, if there is little happy convergence between the substance of the consensus and the agency's institutional preferences.

16. Even at relatively high levels of consensus, bureaucratic segmentation (or low sectoral autonomy) can dilute the clarity of the boundaries of the win-set. The risks of involuntary defection grow— and bargaining positions weaken—but the rising ability to use domestic dissent as a bargaining chip can offset this effect.

17. Knowledge about the bargaining parties' levels of macropolitical consensus and lateral autonomy can help estimate (a) the identity of the bargaining agents, (b) their relative bargaining advantages, and (c) the expected transaction costs involved in a given bargaining situation. This conceptualization of the bargaining context applies

across regime types and helps transcend the difficulties of defining strong and weak states as a determinant of bargaining outcomes.

On the outcomes of decision making and bargaining over technology:

18. Although external constraints provide an important backdrop against which domestic choices are made, they do not necessarily determine those choices. Recipients are not locked into technological decisions—forced by their international position—even in "most likely cases" involving high technology.[77] Overall, the "obsolescing bargain"—and host states' ability to undo some of the initial unfavorable arrangements—appears to hold for nuclear industries as well. The injunctions of the international nonproliferation regime had only a marginal impact on the industrial and technical characteristics of the Brazilian and Argentine nuclear programs.

19. The direct effects of macropolitical constraints and bureaucratic segmentation can be gauged in a country's performance regarding technology transfer, coordination with research centers and with financing and fiscal incentives, and efforts to support the technological development of national private firms. In Brazil, a relaxation of technology transfer regulations, delays in the coordination of financing and fiscal incentives, limited efforts to support the technological development of national private firms, and poor coordination with the country's research centers can all be traced to such macropolitical constraints, which a segmented context of implementation only reinforced. In Argentina, CNEA screened technology effectively, advanced professional training, integrated academic and industrial institutions, and aided the private sector technologically, often bearing these technological costs single-handedly—all this thanks to its bureaucratic autonomy sustained over decades of macropolitical turmoil.

20. Reliance on foreign technology does not invariably increase foreign equity or managerial control. Moreover, it is possible to capture technological externalities even from "black boxes."

21. Despite recurrent claims to the contrary, particularly during the decision-making stages, the establishment of a nuclear industry in industrializing countries has limited indirect, concatenated effects on the countries' industrial technological infrastructure. In fact, nuclear industries can siphon resources from other technological priorities with greater social overhead value.

A Road Map

Chapter 2 weaves the general argument outlined here with the empirical record by analyzing the sources and impact of differences in macropolitical consensus and lateral autonomy in the nuclear sector. Once I establish how domestic politics shape policy positions and influence the nature of international agreements, I then examine more closely in Part II the impact of differences in consensus and autonomy on state relations vis-à-vis private entrepreneurs (Chapter 3), international bargaining (Chapter 4), and the scientific community (Chapter 5). The conceptual orientation of Part II may be of greater relevance to readers interested in the nature of the state, the political power of private actors, ruling coalitions, the process of international bargaining, and the political relations between scientific communities and the state.

Part III provides a comprehensive analysis of the outcomes of state intervention in nuclear industries. Chapter 6 focuses on direct effects of technology transfer and R&D efforts and Chapter 7 on indirect effects, including technological spin-offs and entrepreneurial backward and forward linkages. These chapters are relevant to conceptual debates about the nature of technological change in industrializing countries, yet they also provide concrete, practical, policy-relevant lessons for practitioners and decision makers concerned with technological modernization of the industrial structure. I cast the discussion of nuclear exports by Argentina and Brazil—two important players in that international market—mostly in the context of learning, linkages, and spin-offs. But there are unintended strategic implications of interest to a wider community concerned with the intersection between nuclear markets and international security.

In the final chapter I discuss the implications of the book's conceptual framework and empirical findings for understanding other industrial sectors, technological choices, and bargaining behavior. The conclusions distill some lessons from this study for other industrializing states considering the establishment of nuclear industries.

CHAPTER 2

Domestic Structures, Institutions, and Industrial Choices

This chapter describes and explains the differences between the Brazilian and Argentine nuclear industries in their emphasis on private versus public, and domestic versus foreign, participation. It then extends the general argument about the sources of industrial policy to explain technology choice. The analysis relates to the program's initial decisions and early phases. The chapter then explores the applicability of the argument to other historical periods. It ends with an evaluation of the usefulness and limits of the structural-institutional perspective advanced here for understanding industrial policy and technology choices.

Alternative Nuclear Paths

In 1971 Brazil acquired its first nuclear power plant from Westinghouse (Angra 1; 626 MWe) through a turnkey transfer not designed to set up a national nuclear industrial infrastructure. Only in 1974–75 did Brazil embark on a large-scale program to that effect, by signing an agreement with the West German firm Kraftwerk Union (KWU), a subsidiary of Siemens, to secure the transfer of eight pressurized water reactors and complete full-cycle technology. Only the first of these eight reactors (Angra 2; 1,245 MWe) was under construction in 1994 and was expected to become operational at the end of the decade; the construction of a second reactor (Angra 3; 1,245 MWe) was discontinued in 1993. Argentina's efforts to establish a nuclear industry started earlier, in the 1960s, with fuel-cycle activities and the acquisition of a first power plant (Atucha 1; 367 MWe) in 1968 from Siemens. Atomic Energy of Canada supplied a second plant (Embalse; 648 MWe) in 1973, and Kraftwerk Union (Siemens) a third (Atucha 2, 745 megawatts, expected to operate in 1997) in 1979.

Table 1 compares Brazil and Argentina according to the respective participation shares of national private, state, and foreign firms in the first three power plants. Participation is disaggregated into supplies of electromechanical equipment and engineering services. These two categories account for about 70 percent and 8 percent, respectively, of the total costs of each power plant (with civil construction and assembly operations accounting for the balance). There is a significant contrast between the contribution of Argentine private firms in electromechanical equipment (13 percent of total electromechanical supplies) and that of Brazil's private firms (less than 2 percent) regarding the first plant. The pattern of greater participation by Argentine private producers in this area is maintained for the next two plants.

In the area of engineering services, the contribution of private firms to the first plant was, in each case, imperceptible. For the second plant, however, the contribution of Argentina's private engineering firms is over three times that of their Brazilian counterparts. Instead, Brazil's newly created state enterprise NUCLEN (Nuclebrás Engineering, a subsidiary of Nuclebrás) provided the bulk of national engineering services.[1] The planned contributions by private engineering firms level off for the third plant, which, in the case of Brazil, never went beyond the planning stage. Argentina contracted for this (third) plant in 1979, under changing political and institutional conditions that I will analyze subsequently. These changes explain, for instance, the creation of a joint (state-foreign) venture in reactor engineering.

Taking a more comprehensive look at the nuclear industry as a whole, beyond the construction of power plants, Table 2 highlights Argentina's broad commitment to integrate national private firms into its nuclear industrial activities. Private companies were engaged in uranium mining and yellowcake production, and in the design and production of power plant equipment, including heavy reactor components, and instrumentation and control inputs.[2] In joint ventures with CNEA, such as Conuar, private firms retained majority ownership.[3] There were no private Brazilian companies in either fuel-cycle

TABLE I

Share of Supplies for the First Three Nuclear Reactors in Brazil
and Argentina by Private, State, and Foreign Firms, in Percentages

| | Plant 1 | | | | | | Plant 2 | | | | | | Plant 3 | | | | | |
| | Brazil Angra 1 | | | Argentina Atucha 1 | | | Brazil Angra 2 | | | Argentina Embalse | | | Brazil Iguape | | | Argentina Atucha 2 | | |
Type of inputs	P	S	F	P	S	F	P	S	F	P	S	F	P	S	F	P	S	F
Electromechanical equipment	<2	—	98	13	—	87	27	—	73	40	—	60	28	20	52	50	—	50
Engineering services	<2	—	98	<2	—	98	10	30	60	35	10	55	30	30	40	30	60	10
National participation as a percentage of total direct costs	←8→		92	←38→		62	←30^a→		70^a	←60→		40	←30^a→		70^a	←70→		30

SOURCES: Federal Senate of Brazil, *Relatório de Comissão Parlamentar de Inquérito do Senado Federal sobre o Acordo Nuclear do Brasil com a Republica Federal da Alemanha* (Transcript of the Parliamentary Investigating Committee of the Federal Senate Regarding the Nuclear Agreements Between Brazil and the Federal Republic of Germany, hereafter cited as *Relatório*), vol. 3, pp. 118 and 132; vol. 4 (Brasília; Senado Federal, 1983), p. 81; Federal Senate of Brazil, *Relatório*, vol. 6, book 5 (Brasília: Senado Federal, 1984), p. 324; Sara V. de Tanis and Jorge Kittl, *Twenty Years of Research and Development* (in Spanish) (Buenos Aires: National Atomic Energy Commission, 1976), p. 21; Sara V. de Tanis, *Development of Industrial Suppliers for Argentina's Nuclear Industry* (in Spanish) (Buenos Aires: National Atomic Energy Agency, 1985), pp. 15–16; Jorge Cosentino, "The 'Unbundling' Experience in the Argentine Nuclear Reactor and Power Plant Program" (in Spanish), presented at an international seminar on the unbundling of investment programs in the state sector in developing countries, Buenos Aires, November 1984; *Clarín*, April 15, 1984, p. 14; *Nuclear Engineering International* Feb. 1989, p. 53 (Reprinted with revisions from *International Organization* 47, no. 2 [Spring 1993]: 207–34.)

ABBREVIATIONS: P = private domestic firms; S = state firms; F = foreign firms. In Brazil's plant 1 a single firm accounted for more than 90 percent of total electromechanical equipment and engineering supplies.

^{a}Projected in 1975.

activities or heavy components production. Instead, the state holding firm Nuclebrás created joint ventures with foreign partners in most nuclear activities. The minimal participation of the domestic private sector in Brazil can be easily detected in the empty cells of Table 2 (under "private").

These preliminary observations point to a consistently higher level of participation by Argentina's private firms when compared to Brazil's. The lower participation of Brazil's private sector allowed Brazil's state firms and their foreign partners a more extensive role, as indicated in Table 1. Looking at these differences in absolute terms, however, obscures a more profound contrast regarding the role of national private firms. A more valid measure of each country's commitment to maximize the role of such firms can be found by comparing their actual contribution shares (shown in Table 1) with their respective available endowments, that is, against the capacity of private firms to supply components and services, regardless of cost considerations. We can thus gauge state efforts to involve a maximum number of private firms as a function of: attempts to probe their extant capacity, willingness to tolerate higher costs (relative to imported equivalents), and state efforts to absorb private investments in new machinery, training, and quality assurance. From this perspective, Argentina's commitment becomes even more significant.

TABLE 2

Differences Between Brazil's and Argentina's Nuclear Industries (1964–1986)
(Foreign, State, and National Private Participation in Nuclear Industrial Activities)

Nuclear industrial activity[a]	Brazil (Technology: Light water/ enriched uranium)			Argentina (Technology: heavy water/ natural uranium)		
	Foreign	State	Private	Foreign	State	Private
Uranium exploration, mining, and yellowcake production	(1) joint venture (NUCLAM)	(2) Nuclebrás (NUCLEMON) (3) Nuclebrás (Poços)			CNEA (Malargüe)	Nuclear Mendoza Sanchez-Granel (Los Gigantes)
Conversion UF$_6$ or UO$_2$		Nuclebrás			CNEA	Nuclear Mendoza
U/enrichment HW production		joint venture (NUCLEI/NUSTEP)			CNEA (HW)	
Fuel elements		Nuclebrás (Resende)			joint ventures (1) CONUAR (fuel pellets) (2) FAE (zircalloy tubes)	
Reactor engineering		joint venture (NUCLEN)		Siemens AE Canada KWU	CNEA	INVAP
Electro-mechanical components					ALTEC CORATEC Private Firms	
Heavy reactor components		joint venture (NUCLEP)		SIEMENS AE Canada KWU		IMPSA

SOURCES: Relatório Federal Senate, vols. 1–16; Nuclebrás Annual Reports, 1983–86; Tanis, 1986 and 1990.
ABBREVIATIONS: AE = Atomic Energy of Canada; ALTEC (Rio Negro province); CNEA = National Atomic Energy Commission; CONUAR and FAE = majority owned by private group Pérez Companc (PECOM), minority by CNEA; HW = heavy water; IMPSA = Industrias Metalúrgicas Pescarmona; INVAP = Investigaciones Aplicadas (Río Negro province, semi-private); KWU = Kraftwerk Union; Nuclear Mendoza (Mendoza province); NUCLEN = Nuclebrás Engineering; NUCLEP = Nuclebrás Heavy Components; U = uranium; UF$_6$ = uranium hexafluoride; UO$_2$ = uranium dioxide.
[a]Does not include small-scale pilot plants.

Brazil's private capital goods infrastructure in electromechanical and engineering supplies was far more internationally competitive in 1974 than in Argentina in the early 1960s, when the latter's nuclear industry was designed.[4] The custom-made capital goods sector supplying power equipment, in particular, enjoyed more modern production and R&D facilities.[5] Between 1969 and 1975 Brazilian firms had invested substantially in training, new facilities and equipment, and quality control, precisely to meet the potential demand from emerging sectors such as the nuclear program. By the early 1970s Brazil was the second largest producer of capital goods in the developing world (China was the largest), and by 1976 over 90 percent of Brazil's private engineering firms were active internationally. A comprehensive probe of the sector found the industry mature enough to contribute about 54 percent of nuclear power plant inputs immediately, and 70 percent soon after.[6] Private entrepreneurs claimed they could produce most components (except the primary circuit) and could perform about 90 percent of all the engineering tasks.[7]

Yet despite the high potential for private sector participation, these firms were allocated only about 30 percent of electromechanical supplies and far lower percentages of engineering services. The nationalization of inputs for each consecutive nuclear plant was to be achieved mainly through the creation of state firms.[8] Thus, NUCLEP (Nuclebrás Equipamentos Pesados)—which became the largest producer of heavy components for nuclear plants in the industrializing world—and NUCLEN, in the area of reactor engineering, were created in 1974 as subsidiaries of Nuclebrás. NUCLEP and NUCLEN were established as joint ventures with the foreign supplier KWU, displacing domestic private firms from their markets. Although local firms were capable of supplying turbogenerators, their participation in this area was never considered.[9]

Argentina had a far more modest sectoral industrial capacity (mostly of small- and medium-sized firms).[10] Yet Argentina allowed its private firms to contribute the largest share of domestic inputs they could bear; the state—that is, CNEA—also absorbed price differentials (relative to imported counterparts) as well as private firms' training and R&D costs. Even though domestic inputs were at times estimated to be double the costs of their freight-on-board (FOB) equivalents, Argentina's bids for power reactors required prospective suppliers to maximize "national content." Bidders were asked to provide explicit lists with the sources and specifications of all reactor components, so that Argentine negotiators could more readily assess which components might be produced domestically. CNEA conducted several probes of three hundred to six hundred private industrial and engineering firms very early in the process to establish their potential contributions to the nuclear program.

Whereas Brazil expanded state entrepreneurship with NUCLEP and NUCLEN, Argentina worked to enable private firms to upgrade production facilities and design skills. CNEA was guided by the principle of state subsidiarity, that is, to intervene only when the private sector was unable to do so. In fact, private firms were extremely reluctant to enter this market, and CNEA committed itself to reduce the "stategic uncertainty" of private firms by providing a more or less predictable environment against which these firms could invest in new production lines. The main instruments in this overall strategy were the unpackaging and screening of foreign technology and the use of incentives such as credits, tax exemptions, and subsidies.[11]

In sum, state intervention in Argentina was closer to what Samuels (1987) defines as market-creating or conforming, whereas in Brazil it bore the mark of a market-displacing pattern. As argued, an array of common characteristics underlying the economic, industrial, and political evolution of these two countries highlights these differences. The contrast (in emphasis, and relative to their respective endowments) should not obscure an overall strong state presence, as regulator and entrepreneur, in both programs.

Different choices of industrial structure involved different technological paths as well. Brazil opted for light water/enriched uranium technology, readily available in the international market as the leading reactor type. Argentina selected the less commercially and technically desirable heavy water/natural uranium technology. The argument I develop in the next section is subsequently applied to explain the links between industrial strategy and technology choice.

Macropolitical Consensus and Bureaucratic Autonomy in Nuclear Policy

The concepts of macropolitical consensus and bureaucratic autonomy can now be marshaled to make sense of our preceding discussion of alternative nuclear industrial and technological paths in Brazil and Argentina:

In Brazil, a more or less consensual hierarchy of goals and means,

*and a segmented decision making, constrained the range of options
in such a way that nuclear policy followed the core parameters
of Brazil's industrial model: rapid (particularly export-oriented)
growth and macroeconomic stability. State entrepreneurship and
foreign technology had increasingly become the means, leaving less
room for national private industrial and technological resources.* As
we may recall from Figure 1, these conditions characterized cell III,
where conditions of relatively high macropolitical consensus and
low lateral autonomy pointed to the importance of the generic industrial pattern in explaining sectoral policy.

*Argentina's characteristically low macropolitical consensus, and
the tripartite division of the state among the armed services,
strengthened the autonomy of each service within their respective
industrial spheres of influence. These conditions broadened the
range of options available to the Navy's CNEA, and allowed its own
institutional preferences to prevail. These preferences led to the
Commission's special emphasis on domestic private entrepreneurial
and technical resources.* This situation approximates the attributes
of cell II in Figure 1, where a relatively low consensus and high lateral
autonomy insinuate that sectoral institutional interests and trajectories play a particularly important role, offering a useful shortcut to
understanding policy choice.

BRAZIL: THE "MODEL," THE "MIRACLE," AND BUREAUCRATIC SEGMENTATION

The considerable autonomy of state structures—or their ability
to act independently from the power of social classes or interest
groups—strengthened macropolitical consensus in Brazil in the late
1960s and early 1970s. State autonomy in Brazil can be traced to the
absence of a hegemonic class, although it did not imply politically
impotent economic forces.[12] Whether one accepts structural Marxist interpretations of the 1964 military coup or not, the industrial
classes were far more willing to maintain stable support for core state
strategies than was the case in Argentina.[13] Most modern industrial
sectors, national finance, and agribusiness backed the consensus,
which was more likely to benefit them than was the alternative of a
strongly populist, primarily import-substituting industrialization.[14]
The premise that all members of the alliance would benefit from the
accumulation of industrial capital underlied that consensus.[15] As argued earlier, macropolitical objectives rarely amount to a clear and

coherent whole. Yet, relative to the Argentine case, and relative to other historical conjunctures in Brazil itself (since the late 1970s, for instance), there was considerable consensus—within the ruling coalition—over core objectives in Brazil between the late 1960s and mid-1970s.[16]

Rapid growth emphasizing exports (and selective import-substitution) and macroeconomic stability were at the heart of that consensus.[17] The strategy was often described as the basis of Brazil's economic "miracle" (a 10 percent annual rate of growth between 1967 and 1973) and was designed to provide the military with a legitimating basis for its political control of the state.[18] The 1974 oil crisis endangered these core objectives; Brazil's dependence on external sources of energy for over 80 percent of its domestic consumption threatened its balance of payments and its energy-intensive path to industrialization.[19] Thus, the nuclear industry was designed to address this broader political objective: the continuation of the consensual model of industrialization.[20] State firms and readily available foreign capital and technology—that is, joint ventures such as NUCLEP—provided the instruments. State entrepreneurship had become, after all, the engine of growth and implied, in no few instances, the displacement of private capital.[21] The state came to control high technology infrastructural and intermediate goods characterized by low returns, high risk, and long gestation—such as electricity, gas, oil, telecommunications, iron ore, shipping, and steel—that are basic inputs to modern private industry.[22] Major beneficiaries of expanded electrical generation capabilities were to be the (mostly public) metallurgic sector (the largest single industrial user of energy), chemical and petrochemical industries, paper, and users of intermediate products, including many private producers of mechanical and electrical machinery.[23] Thus, the scale, technical features, and political-economic characteristics of the nuclear agreements with West Germany and KWU were compatible with the broader objectives and instruments of Brazil's industrial policy at that time. Foreign financing was a key feature of that strategy, particularly under Finance Minister Delfim Netto, even where it implied higher ratios of suppliers' credits over regular loans. Suppliers' credits often increased the shares of foreign equipment, and KWU's handsome offers of suppliers' credits were designed precisely to maximize its own supplies and those of its associated German firms.[24]

The consensus regarding core objectives was enforced by the main economic bureaucracies and permeated the internal operation of sectoral agencies like Nuclebrás and the National Nuclear Energy Commission (Comissão Nacional de Energia Nuclear—CNEN), thus constraining their options. The centralization of decision making, designed to check the autonomy of state enterprises, increased significantly in the aftermath of the 1973 oil crisis and was at its height during 1974–75, when the nuclear agreements were negotiated.[25] The Economic Development Council (CDE), headed by President Geisel himself, was the source of general policy orientations guiding the state bureaucracy as a whole.[26] CNEN's bureaucratic autonomy withered earlier in 1967, with its transfer from direct subordination to the presidency to effective accountability to the Ministry of Mines and Energy. CNEN thus played a marginal role in designing the nuclear sector and in negotiating the agreements with KWU.[27]

The main economic bureaucracies—guardians of the model and the "miracle"—defined the parameters of industrial policy in general, and of the nuclear sector in particular.[28] Brazil's nuclear decisions of 1974 were made by the top executive in a horizontally segmented context where nuclear sectoral institutions had little autonomy.[29] The architects of the nuclear industrial program were President Geisel (former director of Petrobrás, the state's oil monopoly), General Golbery de Couto e Silva (military ideologue of Brazil's "model" of economic development), Paulo Nogueira Batista (director of the state enterprise Nuclebrás and former chief of the Foreign Ministry's Economic Department), Foreign Minister Azeredo de Silveira (architect of the "economic miracle"'s foreign policy), Energy Minister Shigeaki Ueki (former director of Petrobrás), and General Hugo Abreu (Chief of Military Cabinet of the Presidency and secretary-general of the National Security Council).

Implementation was in the hands of Nuclebrás, the state nuclear sectoral holding company, formally accountable to the Ministry of Mines and Energy. FURNAS (a subsidiary of Eletrobrás, the state utility responsible for the construction and operation of nuclear plants) was the client. Partial financing was the responsibility of the National Economic Development Bank (BNDE until 1982, later BNDES) and other agents. Finally, licensing, regulation, and effective R&D were more the domain of CNEN than of Nuclebrás's R&D Center. This lateral segmentation did not exist in Argentina, where all research, international bargaining, licensing, and financing functions were CNEA's responsibility.

The dominance of central economic and energy agencies in Brazil curtailed radical departures from core macropolitical objectives. Brazilian decision makers regarded the possibility of slowing down the development of a nuclear industry—by fully integrating private Brazilian firms at the outset—as just such a departure.[30] Allowing these firms to provide a higher share of equipment involved risks of energy undersupply, delays, and capital and technology shortages.[31] This solution was, therefore, outside the domestic win-set that guided Brazil's bargaining with West Germany and KWU.

This interpretation of the structuring of Brazil's nuclear industry challenges the conventional wisdom regarding the role of the military in what could be intuitively considered its logical industrial fiefdom. In the functional division of labor between the civilian and military segments of the technocracy, civilians appeared to have played a predominant role in shaping industrial policy.[32] The nuclear program's technical and industrial characteristics, therefore, were the domain of central economic agencies. The lateral dominance of central state agencies was reinforced by an extensive and intrusive military intelligence network—the Serviço Nacional de Informações (SNI), accountable only to the president—that helped maintain a standardized outlook among top- and middle-level technocrats.[33] Such prerogatives in the hands of the civilian technocracy reduced the military's leverage in selecting a particular nuclear technical option.[34] On the whole, however, the military embraced the large-scale nuclear program of 1974 wholeheartedly, for it perceived the program's scale and scope as allowing Brazil to burn stages in the development of a nuclear industry.[35]

ARGENTINA: MACROPOLITICAL CHAOS AND THE "MAVERICK"

There is widespread agreement on the fundamental absence of macropolitical consensus over industrial policy in Argentina since 1955, although scholars have differed in their explanations for such a low consensus.[36] A leading argument, for instance, advances that the autonomy of the Argentine state was consistently challenged by political and economic forces—agroexporting sectors and industrial entrepreneurs—that precluded the state from consolidating a stable industrialization strategy.[37] Potential partners to a ruling coalition saw their interests better served by exercising a veto power and by providing an erratic, selective support of policies, rather than by effectively throwing their lot into a stable alliance. The military regimes of 1967 (the "Argentine Revolution") and of 1976–81 (the "New Order"), in particular, attempted to launch new economic programs that constituted a major break with past import-substitution and that were likely to attract modern financial, large-scale industrial, and export-oriented sectors. Yet the position of Argentine industrialists and Pampean producers was as contradictory and elusive as that of a firefly: at times either or both were "inside the tent"; at other times they were not.[38] Their hesitancy is sometimes traced to their short-term view of profitability, their rejection of statism, and the deterring effect of a tight alliance with a conflict-ridden, ineffective military institution in control of the political process.[39]

These conditions narrowed the set of converging objectives among potential partners to a ruling coalition and precluded the emergence of a strong and durable consensus.[40] As a result, Argentine coalitions were not only highly unstable, but their shifting boundaries were much harder to identify than was the case for Brazil in the late 1960s and early 1970s. The exclusion of certain political forces with divergent interests (small-scale producers of standardized products, labor, political parties) from the coalition was designed to strengthen the basis of consensus. These forces, however, proved too powerful to be discounted altogether, as the intermittent assaults by the Confederación General Económica and the Cordobazo revealed.[41] The attempt to genuflect to nonmembers or occasional members of the coalition, in exchange for political support, weakened the consensus even further. Tensions within the armed forces, stemming from diverging institutional interests, contributed to weaken macropolitical consensus.[42] Distributional struggles among potential coalition members in Argentina had more of a zero-sum quality than in Brazil, which had a relatively enduring coalition of interests.[43]

Argentina's shifts between attempts at macroeconomic balance and inward-looking policies and their reversal for most of the post-1955 era were a symptom of this feeble consensus.[44] This incoherent pattern of economic and industrial policy, and the resulting turmoil, reinforced factionalism and dissent. Low consensus may also explain both an arresting level of cabinet instability and why a succession of economic "czars" (like Ministers Krieger Vasena and Martínez de Hoz) failed to imbue state agencies with a coherent program.[45] Challenges by the National Development Council and an array of state enterprises, for instance, frequently undermined orthodox policies. The ambiguity in macropolitical objectives expanded the range of options that sectoral agencies could pursue.[46] In this permissible context, agencies with high levels of bureaucratic autonomy could maximize their ability to define and implement their own policy preferences and to expand their jurisdiction. CNEA did just that, when it became the sponsor of selected industrial and engineering firms and of a large elite segment of Argentina's scientific-technical community.

CNEA's autonomy originated with the de facto tripartite division of the Argentine state among the three armed forces, which operated since 1955. In that year, the military ousted President Juan M. Perón, marking the beginning of a cycle of military juntas and brief constitutional interludes. Each service secured centralizing authority over certain industrial sectors under their exclusive jurisdiction.[47] This arrangement reflected a delicate balance of power among the forces. Any decree-laws required consultations with the Military Council, composed of all three services; the Navy, however, gained significant weight in the Council in the aftermath of 1955. The power formula allocated the presidency to the Army, the vice presidency to the Navy, and the remaining three other places in the junta to the Army, Navy, and Air Force, respectively.[48] In practice, the president had no effective control over agencies and programs under the jurisdiction of each service branch.

The Army thus controlled the General Directory of Military Production (DGFM), which coalesced an array of state enterprises in steel, timber, petrochemicals, and electronics. Created in 1941, DGFM controlled fourteen state enterprises by 1945, including the first pig iron factory (Altos Hornos de Zapla) and twenty thousand employees. The state steel company, SOMISA, created in 1947, came under the Army control of General Savio.[49] By the early 1960s, 20 percent of DGFM's production was for military purposes and the remaining 80 percent for civilian use, including supplies for major state firms such as Yacimientos Petrolíferos Fiscales (YPF—oil), Gas del Estado, and Ferrocarriles (railroads), as well as for private industries. It is estimated that DGFM accounted for 5 percent of the country's gross national product (GNP) by the late 1970s, with 22 companies earning $2.2 billion annually through the production of liquid gas piping, oil-drilling machinery, railroad equipment, and petrochemical products.[50]

The Air Force's niche included automobiles, aluminum, agricultural machinery, the National Directorate of State Industries

(DINFIA)—which superseded the former Military Aircraft Industry—and the National Mechanical and Metallurgical Industries (IAME).[51]

The Navy came to control budgetary allocations and major appointments and policy decisions at CNEA and at the National Shipbuilding and Naval Factories (AFNE). Such autonomy allowed an impressive continuity of leadership at CNEA and the vertical functional integration of the nuclear sector under its aegis.[52] This principal-agent relationship enabled the Navy to shape a nuclear industry more compatible with its own institutional interests than with energy requirements as a whole.[53] It also allowed the Navy to maintain—throughout 30 years of political upheaval and low macropolitical consensus—the fundamental technical-industrial characteristics of the nuclear program. These characteristics included an emphasis on procurement of industrial inputs and expertise from domestic private firms and technical groups. Maximizing national inputs was a symbol of technical competence, efficiency, and achievement, which the Navy could deploy in its rivalry with the other two services. The Army and the Air Force had continuously challenged the Navy's institutional power—through fratricide combat in 1962—and succeeded in eroding the position the Navy enjoyed at its heyday in 1955.[54] The possibility of a greater role for national firms and domestic technical resources enabled CNEA to create legitimizing constituencies and clientelistic networks among industrial entrepreneurs and scientific-technical elites.[55]

This strategy fit naturally with the Navy's traditional penchant for technical excellence, often held as an advantage over the Army. The Navy regarded the development of a national nuclear industry as the core asset in its fiefdom: a useful instrument to mobilize productive activities in the electrical, electromechanical, chemical, and metallurgical sectors, which played a fundamental role in the diffusion of technical capabilities and the modernization of industrial production. CNEA's institutional characteristics—meritocratic recruitment, stability, and corporate identity—aided in maintaining technical excellence. These characteristics contrasted with the Army's short-lived control of the nuclear program in the early 1950s. In 1951 Perón announced to the world, prematurely, that Argentina had mastered nuclear fusion, before verifying the fraudulent claim of exiled Austrian physicist Hans Richter, who headed the program. Richter had consistently and successfully challenged the authority of Army

Colonel Enrique González, who was initially in charge of the incipient nuclear program (and a rival of Perón). The Navy took over in 1952, when Admiral Iraolagoitía helped uncover Richter's sham. With Perón's ousting in 1955, CNEA was restructured under Admiral Quihillalt, who headed the agency for eighteen years—until Perón's return from exile in 1973—surviving a succession of eight (civilian and military) governments.[56]

The strategy of engaging private entrepreneurs was also highly compatible with the Navy's classical liberal support for business and state subsidiarity. The Navy was influenced by the British and American models, while the Army embraced a Prussian and statist tradition. Secular groups and political parties linked to the wealthy liberal Anglophile oligarchy supported the Navy and provided its core recruitment pool.[57] Yet Navy support for national private entrepreneurship was no laissez-faire liberalism. True liberals—of the Alsogaray orthodoxy—generally opposed the nuclear program's emphasis on domestic content on grounds of economic efficiency, as did the electrical utilities.[58] The occasional challenges to CNEA's strategy were weakened by the agency's ability to reimburse local suppliers—for all expenses in R&D, training, and technical upgrading, which often doubled the price of imported equivalents—directly out of its budget.

In contrast to Brazil, the primacy of the armed forces' sectoral interests in Argentina constrained the ability of the civilian technocracy to shape industrial policy. The interpenetration of military and economic power was not peculiar to Argentina, but its extent was unique in Latin America.[59] The institutional military control of productive sectors was deeply rooted in Argentina before it gained any significance in Brazil, in the late 1970s.[60] Neither did intraservice rivalry in Brazil ever reach the extremes that it did in Argentina. Brazil's pervasive National Information Service (SNI), a centralized military intelligence apparatus, had no real Argentine counterpart, despite attempts like the "Sistema" (CONASE and CONADE) to combat military factionalism and to coordinate economic policy and national security.[61] Notwithstanding internal divisions within Brazil's armed forces, the military managed to maintain a unifying compromise, particularly during the decade following 1969.[62] Disagreements were submerged "lest they fall into the pattern of the Argentine military whose warring factions are unable to impose any consistent policy."[63]

Industrial Strategy and Technology Choice

The differences in industrial policy analyzed in the preceding section implied differences in technological paths as well. More specifically, the choice of industrial structure involved the adoption of a particular reactor technology and fuel cycle. Light water/enriched uranium cycles were better suited to Brazil's goals of creating a nuclear industry in a relatively short time ("leaping" over stages) and acquiring mastery over the complete fuel cycle under favorable financial conditions. As the leading technology in the global nuclear market, light water reactors ensured reliability, short delivery time, and, in the case of vendor KWU, a willingness to supply complete fuel-cycle technology and financing.[64] In weighing the alternatives— one of which included a more costly and long-term program involving the extensive participation of Brazilian firms—decision makers opted for less expensive and more readily available foreign technology.[65] The risks of light water reactor cycles included perpetuating external fuel dependency, particularly since the enrichment procedure contracted for (the jet-nozzle technique) was not commercially proven. Enriched uranium was a monopoly of the U.S.'s and the Soviet Union's nuclear industries, until the emergence of the European consortia EURENCO and EURODIF. Moreover, because of its potential use in the production of nuclear weapons, the commercialization of enriched uranium had been affected by restrictive conditions, contained in guidelines approved by the major suppliers, through the Zangger Committee, and later, through the "London Suppliers Group."

The arguments in defense of the agreements reached with KWU emphasized their comprehensiveness; the agreements provided for the transfer of the complete nuclear cycle, including enrichment and reprocessing technologies, the diffusion of which, as we have seen, was severely restricted because of the technologies' strategic implications. In other words, effective acquisition of such technologies implied decreased dependence on U.S. enriched uranium supplies, savings in uranium consumption, and a reduction of waste-related problems (through reprocessing). The heavy water alternative was regarded as too costly, whereas the light water option had advantages of lower capital investment and greater operational experience. The purchase of eight similar reactors would ensure standardization, reduce learning costs, and increase safety. Table 3 provides an outline

TABLE 3
Technical Paths and Perceptions of Political Risks and Opportunities

Main risk associated with the technology		Potential costs	Expected opportunities
Brazil	Argentina	Brazil and Argentina	Brazil and Argentina
		HEAVY WATER	
Threat to continuity of 1969–74 economic model		Longer lead time	Greater levels of domestic participation: Entrepreneurial Scientific/technological
		Greater technical uncertainty	
		Lower efficiency	
		Higher cost	Fuel independence (natural uranium)
		LIGHT WATER	
	CNEA's and Navy's strategy of clientelistic inclusion and technical excellence	Short- to medium-term fuel dependence	Leading technology worldwide
		Lower levels of domestic participation: Entrepreneurial Scientific/technological	Lower cost
			Greater technical reliability
			Shorter lead time
			Potential for "extracting" sensitive fuel-cycle technologies from supplier
			Greater financing facilities (suppliers' credits, bank loans)

of technical paths and their associated political risks and opportunities, as perceived by the relevant decision makers in each case.

Argentina's CNEA embraced heavy water reactors because they were presumed to allow greater levels of domestic entrepreneurial and technical participation.[66] First, the production of heavy water was (perhaps wrongly) assumed to be less complex than uranium enrichment. Thus domestic firms could conceivably carry out every stage in the fabrication of fuel elements.[67] Moreover, natural uranium was abundant in Argentina (and Brazil), and heavy water was commercially sold by several countries. These conditions maximized fuel independence and allowed time for a national technology to mature. Paradoxically, both countries mastered uranium enrichment technology in the 1980s, whereas heavy water production in Argentina became problematic. The presumed opportunities offered by heavy water reactors were eventually offset by greater costs, longer lead time, and more technical risks than originally expected.

Yet, when initial technical decisions were reached in the 1960s, heavy water and its promise of a higher national technological content served CNEA and the Navy well, as a symbol of technical achievement and national prowess. These symbols were used by the Navy to preempt attempted Army intrusions in the definition of nuclear technical-political priorities and in budgetary allocations. In 1972, for instance, during negotiations for Argentina's second power plant (Embalse), President-General Lanusse and the Army backed a Westinghouse light water reactor bid, a clear departure from CNEA's historical selection of heavy water cycles.[68] The electrical utilities favored enriched uranium cycles for many of the same reasons that led Brazil's technocrats to embrace that path.[69] Only the Peronist success in the 1973 elections ensured the maintenance of the traditional (heavy water) course.[70] Once again, in 1976 the military administration of General Videla, strongly influenced by the orthodox policies of Finance Minister Martínez de Hoz, imposed an interministerial committee on CNEA, challenging—unsuccessfully—the rationale for Argentine firms' participation in Atucha 2 (the third power plant) and the continued reliance on heavy water reactors.[71]

Further Evidence: Comparative Dynamics

How useful is the argument advanced in this chapter to explain nuclear industrial policy in other historical periods, for both countries? Can different structural and institutional conditions—levels of consensus and autonomy—explain alternative policy outcomes in Brazil during the early 1960s or during the late 1970s?

The early 1960s were a turbulent period in Brazil's political history, characterized by dramatically low macropolitical consensus.[72] The old populist alliance collapsed and a severe macroeconomic crisis made it impossible to govern without alienating one segment or another of potential ruling coalitions.[73] President Jânio Quadros assumed office in 1961, following a long series of failed attempts at economic stabilization, and resigned within seven months, in August. President João Goulart inherited a deepening economic crisis, recession, hyperinflation, and default on international debt. Strong class and sectoral tensions led to increased political mobilization and intense polarization.[74] This was the volatile crisis preceding the April 1964 military coup, which overthrew President Goulart.

CNEN enjoyed a high degree of lateral autonomy during those years. As a federal *autarquia*, with administrative and financial autonomy, and prerogatives to negotiate treaties, financing, and sale of feasible materials, CNEN was directly subordinated to the presidency, much as its Argentine counterpart, CNEA. Under President Quadros, CNEN advocated national research and control of nuclear resources, diversification of external sources of technology, fuel independence, and active domestic procurement of nuclear reactor components among private firms. Heavy water and thorium cycles, which had strong support among scientists and technologists, were the natural technical choice for such an industrial blueprint.[75] In 1961 CNEN signed an agreement with private industrial entrepreneurs to promote production and trade in nuclear minerals, and created the mixed (state-national private) firm COMANBRA (Companhia de Materiais Nucleares de Brasil). The resignation of Quadros in 1961 led to the cancellation of a project to launch an international bid for a natural uranium/heavy water reactor, with significant participation by Brazil's national industry.[76] Yet President Goulart recommended in 1963, shortly before being deposed by the military in 1964, the construction of a nuclear power plant with those same technical-industrial characteristics. There is a striking parallel between Brazil's nuclear choices during those years and those of Argentina. In both cases nuclear policy was defined under conditions of low macropolitical consensus and high sectoral institutional autonomy, with the nuclear agency actively seeking the inclusion of private industrial interests and technological networks.

This policy was completely reversed following the 1964 coup. As argued earlier, by the late 1960s macropolitical consensus and a segmented decision making led to a program harmonious with an economic strategy emphasizing state entrepreneurship and foreign technology. The political logic of the model of industrialization led to the 1974–75 agreements with West Germany and KWU for a large-scale industrial program. The new directives on national nuclear energy policy since 1967 concentrated the implementation of energy policy in the Ministry of Mines and Energy. CNEN was no longer subordinated to Brazil's presidency, but to the Minister of Mines and Energy, and its mission reflected the regime's macropolitical objectives.[77] Its technical expression was the adoption of light water fuel cycles and the termination of all other projects (specifically those based on natural uranium and thorium). By 1971 representatives from an interministerial energy commission, and from FURNAS, Eletrobrás, and CNEN, recommended to the Minister of Mines and Energy the turnkey purchase (with minimal national input) of a Westinghouse enriched-uranium pressurized water reactor (PWR), as Brazil's first nuclear power plant (Angra 1).[78]

Changes in macropolitical consensus and lateral autonomy by the late 1970s allow us one other test of the usefulness of this framework in explaining policy over time. At this historical juncture the declining performance of the political-economic model of industrialization, and the rising political power of industrial entrepreneurs, eroded the old consensus. The economic crisis of the mid-1970s sharpened the contradictions between maintaining macroeconomic stability and strengthening the position of local private capitalists.[79] In a much publicized statement entitled "Manifesto of the Eight," major industrial leaders formerly supportive of the ruling coalition made their dissatisfaction with the regime—and with expanding state entrepreneurship—evident.[80] Their continuous support was essential for a military regime with little popular legitimacy. The nuclear agreements negotiated five years earlier could have hardly been reached at this time. While macropolitical consensus was waning, the Navy strengthened its control over CNEN, reinvigorating the Commission's autonomy and restoring many of the prerogatives it enjoyed in earlier times (under Goulart). Partnerships with private industrial and technological networks revived technical options compatible with those interests. By 1988 Nuclebrás was dissolved, and the joint (state-foreign) venture NUCLEP was privatized. The fabric of civil-military relations in the postauthoritarian era in Brazil—where the armed forces maintained certain controls in exchange for remaining aloof from direct political intervention—reinforced CNEN's lateral autonomy, this time under the Navy's control, in the 1980s.[81]

A dynamic analysis of the Argentine case is possible only if we address the postauthoritarian period, in the latter part of the 1980s. Until then, low macropolitical consensus and the nuclear agency's high lateral autonomy were a constant, which allowed CNEA to pursue a remarkably consistent policy for almost 30 years. Without such autonomy the program would have been more influenced by other bureaucratic forces. The economic ministries and the electrical utilities would have pushed for a more "rational" (less expensive) program, with lower levels of national participation, maximum imported equipment and, most probably, light water reactors.[82] The Army might have injected its preferences for statist alternatives, as it did with the industrial complex managed by Fabricaciones Militares (DGFM).[83] In short, in the absence of macropolitical consensus, conflicting bureaucratic pressures would have increased the range of possible options and, most likely, the program's incoherence (these conditions characterize cell IV in Figure 1).

Macropolitical consensus remained fundamentally fragile in Argentina until 1990, when President Carlos S. Menem introduced an orthodox stabilization program, backed by a broad spectrum of industrial interests and even fractions of labor.[84] Expressing the new consensus, the Planning Secretariat engineered in 1991 the reorganization of CNEA and the privatization of associated firms. By then, CNEA's autonomy had been drastically curtailed, following the collapse—with the advent of democracy after 1983—of the old tripartite arrangement among the armed forces. Other state agencies, including the electrical utilities, were now active in nuclear policy. A segmented decision making allowed consideration of alternative technical cycles, which CNEA's lateral autonomy had prevented for so many years. The old policy and partnerships gave way to a new—leaner—program, which began exploring the possibility of privatizing power plants and turning over the third plant (Atucha 2) from Argentine control to Siemens, for completion as a turnkey. Quite a new beginning for the Argentine nuclear industry!

This exercise in comparative dynamics raises the issue of mutual influence between evolving state-society relations and institutional changes on the one hand, and sectoral policy on the other. At time T, macropolitical consensus and a segmented decision making can lead to an emphasis on state entrepreneurship and joint ventures with foreign technology suppliers. At time T_{+1} such policy outcome affects (reinforces or transforms) the structural and institutional arrangements that shaped it in the first place. The relative success of the technological path adopted by Argentina's CNEA—which helped maintain, for many years, the agency's fundamental characteristics—is an example of outcomes that reinforce such institutional arrangements. In contrast, the creation of the state firm NUCLEP—in what the private sector regarded as an instance of market-displacing behavior by the state—accelerated the consolidation of economic and political forces against *estatização* (statization or overexpansion of the state sector) in Brazil. It thus contributed to the transformation of earlier structures, which allowed NUCLEP to come into being in the first place. Similarly, CNEA's fostering of what was initially a

feeble sectoral entrepreneurial force in Argentina brought about unprecedented pressures by a growing nuclear lobby on a less autonomous state in the 1980s. Finally, sectoral segmentation in Brazil left
policy implementation in the hands of the Planning Secretariat, Eletrobrás, BNDES, and other agencies, a fact that placed heavy budgetary and operational constraints on Nuclebrás. As a result, the program—as originally envisaged in 1974—collapsed incrementally.
The disarray of the nuclear sector, in turn, led to demands for
greater lateral autonomy. I discuss these processes in more detail in
Chapter 3.

Structures, Institutions, and Industrial Policy

The explanation I advance for the choices that states make in defining industrial policy across sectors is inspired by the attempt to
improve our understanding of structural and institutional influences
on industrial policy. While emphasizing structural determinants of
macropolitical consensus, this framework does not fully explore how
different power structures lead to varying levels of consensus. In
other words, explaining consensus is, for the most part, exogenous to
the main argument, even though I provide some suggestions as to
how it may come about. In contrast, explaining the impact of consensus on the definition of property relations among state and private
(foreign and national) actors is a central task here. I find this to be a
more promising research strategy than merely probing the "desires"
of state agencies. As Barzelay (1986:111) argues, the question is not
merely whether separate agencies represent "different normative
visions of the state," as they tend to do, but what structural and
institutional arrangements allow them to pursue their goals and
implement their policies. In other words, "systems are not all-
determining, but units . . . do not choose in a vacuum."[85]

Far from suggesting a "hyper-structuralist trap" (Jacobsen and Hofhansel, 1984) imposing a certain kind of program and not another, I
discuss these conditions as important in explaining constraints and
incentives, although at times insufficient in describing choices and
outcomes. As Gourevitch (1978) argued, "The impact of structures
lies not in some inherent, self-contained quality, but rather in the
way a given structure at specific historical moments helps one set of
opinions prevail over another." The nature and degree of macropolitical consensus—a reflection of structural relations of power em-

bedded in ruling coalitions—and the bureaucratic autonomy of relevant institutions over time and across sectors allow different sets of options to prevail.

The question then arises, where do the options come from? Some are defined by the *content* of the consensus and reflect the shared preferences of the ruling coalition. Whether these options matter, and by how much, is often a function of the *degree* of consensus. High levels of consensus are often accompanied by a more or less clear definition of options that, in turn, will weigh more heavily on decisions. Options also emanate from state agencies that, for the most part, evaluate different industrial trajectories according to their relative impact on the agency's institutional interests. Thus, the analysis of CNEA as a purposive entity found the Navy's nuclear agency embracing a technical-industrial option aimed at maximizing its standing vis-à-vis the Army and the Air Force. That option relied on the mobilization of clientelistic support from private industrialists and scientific-technological groups.[86] CNEA's preferences derived, to a large extent, from the architecture of the Argentine state and its tripartite division of resources among the armed forces. The Navy sought allies in civil society to stem its secular decline in the balance of power within the military triad.[87]

Yet institutional interests are not always easily identified on the basis of an agency's position within the state. First, such position may shift over time so that industrial-technological preferences may be the result of earlier historical conditions.[88] The timing of a program's inception, for instance during a period of institutional creation, can help certain institutional preferences "stick." During the early stages of institutionalization there may be a greater latitude for technological choice. Once rules (formal or informal) are in place, decisions may be constrained by iterative processes. Institutional crises, instead, can provide new opportunities and shift the sector's characteristics in new directions.[89] Second, the identification of the Navy's (and CNEA's) orientation toward state subsidiarity opens up the possibility that ideas and convictions matter, that is, they can affect choices. Hardly a contentious proposition in itself, its value is contingent on the ability to specify *under what conditions* that is the case. Following North (1990:40), ideas appear to matter more when their "price" is low—in our context, when agencies are not required to sacrifice core values such as power, budgets, and recognition. In effect, CNEA's subsidiarity approach and its protection of national private firms were not only relatively costless; they were also highly compatible with the agency's more "objective" preferences (interests) defined previously. This, of course, undermines our ability to gauge the effective impact of ideological considerations.

Finally, we return to the primacy of institutional context. Ideas similar to those of Argentina's CNEA were shared historically by Brazil's CNEN; yet the existence of those ideas was not enough to overrule CNEN's institutional interests in the context of a segmented decision making and a high macropolitical consensus. For CNEN, the costs of acting on those ideas were much lower in the early 1960s; at that time, low macropolitical consensus and high sectoral autonomy enabled CNEN's preferences for heavy water (and thorium) paths, and for a greater role for national private firms, to prevail. Thus ideological considerations can be brought in, without reducing the explanation to them.

Summing up, the argument I advance here: (1) proposes a set of structural and institutional variables that can explain policy over time; (2) helps us identify the conditions under which sectoral institutions matter; (3) offers a guide to understanding how state agencies formulate industrial options and preferences; and (4) explains how the structural and institutional context within which such agencies operate influences bargaining processes with foreign actors.

Having clarified the role of institutional constraints in policy making, I will now explore how private firms respond to such constraints. As I argue next, industrial entrepreneurs do not merely react to inducements and constraints, they are also a key ingredient in the degree and maintenance of macropolitical consensus.

PART TWO

The Process

CHAPTER 3

State and Private Entrepreneurs: Partners or Foes?

Private sector preferences play a key role in shaping macro-political consensus, whether or not the private sector is effectively included in the decision-making process. The ability to formulate and advance a coherent industrial program hinges on the degree to which most critical components of a ruling coalition have converging interests. In this chapter I examine the place of Brazil's industrial entrepreneurs within the ruling coalition's consensus of the post-1969 era and their role in weakening that consensus toward the late 1970s. The nuclear program played a central part in the industrialists' attempt to challenge the model that led to the 1974–75 nuclear agreements with KWU. I then explore how political and institutional constraints and incentives shaped the private microeconomic calculus of risks and opportunities regarding the nuclear program. Next I offer a contrasting evaluation of state inducements and entrepreneurial responses in Argentina. This overview suggests that the segmented decision-making process in Brazil had strikingly different results from the one dominated by an autonomous agency in Argentina. I conclude by presenting the implications of these cases for understanding state-society relations and the behavior of private entrepreneurs in industrializing contexts.

Brazil: Capital Goods, Macropolitical Consensus, and "The Miracle" Imperiled

I argued earlier that macropolitical consensus over industrial priorities was higher in Brazil in the first decade following the military takeover in 1964 than in Argentina for most of the post-1955 period. Although not necessarily formal partners in the policy process, Brazilian industrialists contributed to the high macropolitical consensus characteristic of the post-1964—and particularly post-1969—economic and industrial policies. Industrialists provided a key basis of support for the regime's political-economic model.[1] The junior status of private local firms in the "triple alliance" of state, multinational, and private local capital, characteristic of these years, did not prevent an unprecedented growth in the sector's economic and political power.[2]

Capital goods producers were an elite segment among national entrepreneurs.[3] As a capital goods industry par excellence, the development of a nuclear reactor industry offered considerable opportunities.[4] A review of the evolution of capital goods production since 1964 highlights the benefits of the overall industrial strategy for these firms and helps explain the sector's support for that model of industrialization.

The capital goods sector as a whole was generally assumed to have played a special role in the economic strategy of the regime installed in 1964. In particular, domestic production of capital goods was regarded as a means to increase an indigenous technological capacity.[5] This was particularly the case for custom-made capital goods, which often involve higher degrees of technological effort. Moreover, local production of these goods had been traditionally associated with political windfalls, such as independence from external political decisions and economic cycles.[6] As direct purchaser and financing agent, the state had played a critical role in the development of this sector. Two-thirds of total invoicing in custom-made equipment was procured by state agencies. Private firms accounted for an important share of capital goods output. By the 1970s, Brazil became the second largest exporter of capital goods in the industrializing world (after the People's Republic of China).

A period of accelerated growth in capital goods production began in the mid-1950s, under President Kubitschek.[7] Between 1966 and 1973, the sector's participation in industrial growth grew from 8.8 to

27.8 percent.[8] At the same time, stimulated by fiscal incentives, tariff exemptions, and tied-in foreign credit, imports of capital goods grew by 24.6 percent yearly after 1970, accounting for 42 percent of total imports. This process was part of a strategy, particularly until the mid-1970s, of maximizing short-term growth based on durable consumer goods, which favored imports of capital goods.

Capital goods replaced consumer durables as the major source of industrial growth in the early 1970s, expanding production at an annual rate of 19.3 percent.[9] By the mid-1970s, in the aftermath of the energy crisis, there was a transition to a policy of import substitution and of strengthening local capital goods production.[10] Legislation aimed at protecting this sector increased after 1975, stressing the favorable effects of import substitution in capital goods on a deteriorating balance of trade.[11] The sector became a major priority in the Second National Development Plan adopted in 1974, a plan warmly received by private entrepreneurs. Government agencies were directed to turn more and more to locally owned firms and less to foreign subsidiaries.[12]

Yet sectoral protection in principle was often suspended by other measures, which, in fact, accorded preferential exchange-rate treatment to imports. Public sector expansion increasingly required deficit finance and reliance on external resources.[13] These policy inconsistencies reflected a fundamental ambivalence on the part of the regime regarding the value of a private domestic capital goods industry. State enterprises had repeatedly relied on cheap imported equipment as an instrument for promoting industrialization. Restricting imports, particularly those supplied with foreign credits, was expected to have a detrimental effect on capital formation and growth.[14]

Three main factors encouraged a continued reliance on imports. First, there were serious constraints on domestic financing after 1973.[15] Efforts to increase the domestic share of capital goods supplies after 1974 were thus impaired by the inability to finance large governmental programs from domestic sources. Balance of payment difficulties strengthened the hand of external financing schemes for imported equipment.[16] Supplier credits were easy to obtain in the international financial environment of the mid-1970s. Second, foreign equipment solved the problem of actual and perceived deficiencies in the availability and reliability of certain material and human inputs (components, materials, processes, managerial and technical expertise). By invoking technical requirements and specifications, state managers were better able to justify their overlooking of policy directives protecting local producers, particularly in the area of custom-made capital goods. It should be noted that although local supply of capital goods increased remarkably between 1973 and 1977, the proportion of custom-made capital goods (such as those required for nuclear plants and facilities) over total capital goods imports *grew* after 1975.[17] Finally, imports were promoted to maintain a favorable relationship with foreign suppliers, important partners in the maintenance of the model of industrialization on which Brazil embarked in the 1960s.[18] These considerations—financial, technical, and political—explain the strength of disincentives to protect domestic capital goods.

Summing up, Brazil had an internationally competitive private infrastructure in heavy mechanical and engineering goods and services. Yet there was a discretionary and capricious application of protection, mostly where local production of capital goods would not impair the economic "miracle." Joint ventures with foreign partners were attractive for solving potential financial and technological bottlenecks, while accommodating state entrepreneurship. Foreign loans and suppliers' credits reinforced reliance on imports.[19] Thus, as in various other industrial sectors, the Industrial Development Council (CDI)—the central agency in charge of industrial policy and implementation since 1964—granted tariff exemptions and other incentives to equipment and machinery imported in connection with the nuclear program. Foreign supplies conformed to the liberalizing trade and investment policies of the late 1960s and early 1970s, and to entrenched perceptions regarding the quality and reliability of imported technology and equipment.[20] Imports allowed the accelerated development of a nuclear industry based on ready (and less expensive) technology and on experience, and capable of contributing to the maintenance of the overall energy-intensive strategy of industrialization and growth.[21] At the same time, this reliance on imports also revealed a wavering commitment to domestic producers of capital goods and had the potential for undermining a decade of broad consensus on economic and industrial policy.

Private Entrepreneurs and Nuclear Supplies: Voice and Exit

Despite efforts to pay lip service to the idea of strong private Brazilian participation, the state provided relatively weak sectoral inducements, either through domestic procurement or through more indirect incentives.[22] As we may recall from an earlier discussion, national private firms were allocated a smaller share of supplies than they were able to contribute. Although Brazilian firms claimed the capacity to produce almost all components, with the exception of the primary circuit and multilayered pressure vessels, they were assigned only about 30 percent of electromechanical supplies for the first two plants.[23] A 1974 study by the Bechtel Corporation for the Brazilian Company for Nuclear Technology (CBTN, Nuclebrás's predecessor) found a sample of at least 79 firms able to provide over 50 percent of the total costs of supplies and services in mechanical and electrical equipment, instrumentation and control components, and civil structures and equipment.[24] Pointing to a high level of enthusiasm among these firms, the study estimated that Brazilian participation could increase rapidly to 70 percent.[25]

In the midst of negotiations with the West German firm KWU, CBTN informed the capital goods producers' association (ABDIB) in October 1974 of the possibility of creating a consortium among state, foreign, and local private firms in the nuclear sector, with shares of 51, 25, and 24 percent, respectively.[26] ABDIB's lukewarm reception of this overture stemmed from what it regarded as the proposal's unnecessary requirements for new investments and from the minority status allocated to private Brazilian firms. These firms had been undergoing an expansion that they regarded as sufficient to enable them to meet nuclear requirements, without the need for the additional dedicated investments that state officials imposed as prerequisites for participation.[27] Nuclebrás officials thus justified the relatively marginal role assigned to private domestic firms by arguing that Brazilian firms lacked the commitment to make necessary investments in new machinery, training, and quality assurance programs.[28] Some private firms felt particularly threatened by plans to create a state enterprise capable of competing in the production of pressure vessels, heat exchangers, tanks, and ancillary products.

After CBTN's transformation into Nuclebrás in 1975, contacts with the private sector were held directly with a few major firms, rather than with ABDIB.[29] At that time, NUCLEN (the Nuclebrás subsidiary responsible for power plant engineering) created a department of industrial promotion, later named Department of Industrial Development. Throughout 1975, while negotiations with KWU continued, Nuclebrás Director Paulo N. Batista established preliminary contacts with domestic producers, exploring their participation in two areas: the provision of conventional power plant equipment, starting with the first two plants, and a joint venture (NUCLEP) in the production of heavy nuclear components for the primary circuit, for plants three to eight (see Appendix D).[30] Batista also promised to survey additional firms and provide them with technical assistance and quality control training. Representatives of Brazilian firms in the heavy mechanical, electrical, and electronic area gathered to discuss this proposal, but found it inadequate in its specification of either global investment requirements or their own specific share in the proposed heavy components subsidiary, NUCLEP.

Negotiations with KWU led to agreements establishing specific levels of domestic content. Table 4 provides a breakdown of the share assigned to private Brazilian firms in the supply of electrical, mechanical, and electromechanical components. Two areas, the turbogenerator and special reactor components, were to attain low indices of national participation, even if all (eight) nuclear plants were to be built.[31] For the first two plants, domestic participation would reach 10 percent for the turbogenerator (and zero for heavy and special reactor components), despite an existing capacity by Brazilian firms in the production of turbogenerators.[32] Chapter 4 examines in greater detail the supply-side of heavy electrical equipment for nuclear plants.

In late 1975 Batista announced a plan to place orders with Brazilian firms by early 1976 for $1 billion initially, rising to a $3 billion by 1990, provided all eight nuclear plants were built.[33] Nuclebrás also sought domestic financing through the National Economic Development Bank (BNDE) and the Ministry of Industry and Commerce. Yet these firms expected more concrete signals than public announcements, and they demanded a "market reserve" (a protected niche) that would justify technological and managerial investments. In response, a "Protocol of Market Reserve" was signed in 1976 between Nuclebrás's NUCLEN and three of the largest national manufacturers of components (especially in tanks, vessels, heat exchangers, and cranes), whereby Nuclebrás committed itself to purchase from this consortium at least 50 percent (about $800 million) of *local* supplies for the first four plants.[34] This protocol thus brought the three largest, most modern, and politically influential firms in the sector into the nuclear program, committing them to new investments in quality assurance, training, and equipment. Upon prequalification, they were required by NUCLEN to sign technology transfer

TABLE 4

*Brazil's National Participation in Electromechanical Supplies
(as defined in 1975 accord between Nuclebrás and KWU)*

Components	Plants 1-2	Plant 3	Plant 4	Plants 5-6	Plants 7-8
Turbogenerator	10	15	20	25	30
Heavy components	—	70	100	100	100
Electrical equipment	85	87	90	93	95
Piping	15	20	25	50	65
Instrumentation and control	5	10	60	70	90
Pumps	40	45	47	50	50
Special steel structures	100	100	100	100	100
Heat exchangers	80	90	100	100	100
Ventilation, air conditioners	100	100	100	100	100
Special reactor components	—	10	30	40	50
Rolling bridges	100	100	100	100	100
Valves	10	20	30	40	50
Tanks	90	100	100	100	100
Miscellaneous	70	75	80	85	90
Total share of national nuclear components	30	47	60	65	70

SOURCE: *Relatório* 3 (1984): 132.

agreements with German technology suppliers. However, private Brazilian participation in Nuclebrás's heavy components subsidiary (NUCLEP) was practically dismissed by mid-1976.

We can thus characterize the initial response of private entrepreneurs to the 1974–75 negotiations between Nuclebrás and KWU as one of voicing dissatisfaction with their own overall limited role in the nuclear industry, but not as a confrontational strategy. Some have interpreted this behavior as a function of these firms' preference for traditional products and markets; a sharp opposition to the nuclear program would arguably be unwarranted if their existing markets were to continue expanding (and the nuclear program would, after all, help provide necessary energy requirements for all modern industrial sectors).[35] Private reluctance to invest, however, appears to have been related mostly to the vagaries of the nuclear program, or, in Barzelay's terms, to the "strategic uncertainty" associated with potential policy shifts.[36] The segmented nature of the decision making and implementation contexts contributed to that uncertainty, because it placed different aspects of the program in the hands of agencies with diverging agendas.[37] In 1975 and 1976 official estimates about the likely scope of Brazil's nuclear industry ranged from two to 25 nuclear plants.[38] Moreover, there were strong doubts about Nuclebrás's ability to secure state subsidies and growing concerns about the extent of KWU's influence within Nuclebrás (and therefore, about Nuclebrás's commitment to national firms).[39] The demand by Brazilian entrepreneurs for "market reserve" was geared to minimize political risks (such as a potential reversal in state policies once their own investments were irreversible) as well as macroeconomic uncertainty unrelated to the sector (such as the impact of the economic recession in late 1974 to early 1975 on the availability of credit).[40]

However, the relatively mild initial entrepreneurial response, bordering on acquiescence or on what Hirschman defined as a "high propensity to defer" to normal market constraints and state priorities, rapidly evolved into one of challenge and overt dissent.[41] In the wake of constant delays in actual orders, lack of effective financing arrangements, and indecision regarding tariff and fiscal exemptions for required parts and equipment, Brazilian industrialists became among the most vocal critics of the nuclear program.[42] Despite some relaxation of political controls in the mid-1970s, Brazil was still formally a polity engulfed in military-led authoritarianism. For important segments of the private sector, the gains from macropolitical consensus, and therefore from deference to state tutelage, were decreasing; instead, the opportunities for mobilizing political support in an all-out campaign against *estatização* were on the rise. This much stronger reaction to the nuclear program was in no sense detached from a broader pattern of political realignments. Leaders of major industrial firms critical of the program were also central actors in the broad protest movement by Brazil's civil society against the authoritarian state.[43] Calls for greater economic liberalization were accompanied by demands for political democratization.

The "triple alliance" between multinational, state, and local capital flourished between 1968 and 1973, when the respective gains from an impressive aggregate economic growth muffled internal dissension within the alliance. The benefits from this coalition diminished, in the eyes of the local bourgeoisie, with the economic contraction that followed the 1973 oil crisis. The relative bargaining power of national capital vis-à-vis the state increased, particularly in the capital goods sector, which began challenging prior tripartite arrangements with the state and multinationals, no longer regarding them as variable-sum ventures. National entrepreneurs perceived the state bureaucracy as pursuing separate gains, even while sharing a commitment to capitalism.[44] The benefits from a continued alliance with the state eroded even further following the 1974 elections and the regime's failure to mobilize public support for its program.[45] By 1974, aware of the increasing fragility of the alliance, state officials initiated a dialogue (with CACEX, the foreign trade bureau of the Bank of Brazil) with major private associations in the machine building, electrical, and heavy industrial sectors.

Yet the campaign against *estatização* intensified between 1974 and 1977, with active demands by private entrepreneurs for greater participation in national decisions. Among prominent leaders of the antistatist campaign were major suppliers of nuclear components, such as Claudio Bardella (ex-president of ABDIB) and Luis Eulalio de Vidigal Bueno (owner of Cobrasma and Confab), as well as José Mindlin (director of the São Paulo association FIESP) and Paulo Vellinho (director of the electrical and electronics producers' association ABINEE).[46] Their criticism focused on what they considered an overly centralized decision making, the monopolization of information by state agencies, and the lack of coordination in economic and industrial policy, all of which characterized, in their view, the nuclear program.[47] Leading industrialists joined in signing a manifesto, "In Defense of a Nation Under Threat," which accused President Geisel of *entreguismo* (selling off) to foreign interests.[48]

Efforts by Nuclebrás's NUCLEN, by mid-1977, to increase the role of local firms in the nuclear program made it evident that the national industrial sector was alive and well. Their planned participation was raised only slightly (from 30 percent to 35 percent of all supplies of equipment and services) for the first two plants, but amounted to 93 percent (rather than 70 percent, as originally established) for plants 7 and 8 (see Table 4).[49] Domestic supplies of electronic equipment for the first two plants, originally set at 5 percent, were to increase as well.[50] The net result of negotiations with equipment suppliers was the involvement of about 40 companies in the provision of power plant supplies: 29 in the mechanical sector and 11 in the electrical sector. Nuclebrás's 1982 report points to over one hundred contracts with national industries for equipment and services for the first two plants (Angra 2 and Angra 3), amounting to about $100 million for each plant.[51] The consortium Cobrasma, Confab, and Bardella accounted for about 70 percent of Brazilian mechanical and electromechanical supplies for Angra 2 and 3. Supplies involved mostly thermal and other mechanical equipment and machinery, where national private firms were predominant (unlike heavy electrical machinery).

These general target commitments notwithstanding, formal orders were slow in coming. An employee at one of the firms in the consortium complained that (by 1980!) only one company had received an effective order and a second firm was negotiating another.[53] Delays and financial difficulties began plaguing the program soon after its inception, while a general recession increased the capital goods sector's idle capacity.[54] By the early 1980s, a new assessment of energy requirements, and of the role of nuclear energy, brought about a dramatic contraction of the 1975 nuclear agreements.[55]

Beyond their criticism of the extensive participation of KWU and its associates, the most intense cleavage between private industrialists and state managers developed around what private entrepreneurs regarded as a paradigmatic instance of state encroachment into private markets.[56] The case in point was the establishment of NUCLEP in 1976, as a state-owned joint venture with foreign firms, designed to supply five sets of heavy components for the last six reactors (of the eight included in the agreement with KWU) by 1990.[57] According to Nuclebrás officials, production of heavy components by the private sector proved unfeasible for technical and economic reasons.[58] Therefore, according to the original shareholding agreement between Nuclebrás and KWU/Voest Alpine, Brazilian firms would become minority partners in that joint (state-foreign) venture. Yet none of the 34 invited Brazilian firms regarded NUCLEP as an attractive investment.[59] They argued that the program could have dispensed with it (or at least reduced its scope), allowing private firms to meet that challenge with the mere addition of two or three specialized machines and at a public saving of $500 million.[60] NUCLEP became one of the largest equipment producers in Latin America, with an awesome productive capacity that replicated that of private producers in some cases, with very specialized machinery (some of it not available in the private sector), and with a high level of idle capacity.[61]

Most threatened by the potential production capacity of NUCLEP were the boilermaking, shipbuilding, petrochemical, military, cement, paper, cellulose, and mining industries. This threat led to another protocol in 1980, this time between NUCLEP and ABDIB, requiring NUCLEP to refer orders to firms associated with ABDIB. This agreement reflected the new balance of forces between the state and private industrialists just described. NUCLEP was now required to restrict its activities to the production of heavy equipment for the primary (nuclear) circuit, to place its industrial capacity at the disposal of private firms, and to complement potential deficiencies as a subcontractor for the private sector. These requirements included a commitment not to provide services without prior consultation with ABDIB, so that NUCLEP could perform only those functions that private firms were not capable or willing to undertake. ABDIB was thus able to preempt NUCLEP's entrance into other energy areas, such as offshore platforms and hydroelectrical equipment.[62] A new modus vivendi between state and private entrepreneurship was in place.

Negotiations with private engineering firms resembled in many ways the pattern just described. As in the case of capital goods, the state exercised a predominant influence as a client, particularly for infrastructural projects and processing industries. Most engineering design in Brazil was performed by foreign subsidiaries until 1960, when Petrobrás began providing incentives for local firms. In 1975 the sector was ill-equipped to challenge Minister of Mines and Energy Shigeaki Ueki, who announced that the engineering subsidiary to be established by Nuclebrás (NUCLEN) would not include private firms.[63] NUCLEN was thus formally constituted in December 1975, as another joint venture between Nuclebrás (75 percent) and KWU (25 percent), with a capacity to subcontract with local firms for selected engineering services.

Brazilian firms regarded their share within the joint venture NUCLEN as a regression (quantitatively and qualitatively) from the levels of domestic participation achieved in the construction of Brazil's first nuclear plant, Angra 1, which Westinghouse had supplied as a turnkey in 1972 (prior to the large-scale agreements with KWU).[64] At that time, one firm had been contracted for portions of the electrical, mechanical, instrumentation and control, piping, and structural design. With Angra 2, the first plant in the agreement with KWU, only the last two were contracted out to Brazilian firms, while NUCLEN reserved to itself basic (and even detailed) design on electrical, mechanical, instrumentation, and control systems.[65] With private pressure mounting against *estatização*, in 1977 NUCLEN signed contracts for Angra 2 and 3 with two prominent Brazilian firms, Promon and Engevix.[66] Total contracts with national engineering firms reached about $100 million by 1982. It was not until 1982 that state agencies and enterprises extended a generalized preferential treatment to Brazilian engineering firms. No special legislation enforced reliance on local firms in this area, or provided special credit or fiscal incentives to them; yet a practical preferential policy developed in response to the intense pressures by Brazilian engineering firms.

In the area of uranium mining and processing, stockholding participation was open to private Brazilian firms, but none expressed any interest, at least initially.[67] State firms were dominant in this sector (62 percent of net assets in 1975), whereas private Brazilian firms were mostly in joint ventures, accounting for about 25 percent of the industry.[68] A 100 percent subsidiary of Nuclebrás, NUCLEMON, emerged in early 1976 to retain mining and exploitation of monazite, while encouraging private capital to handle industrial processing of other heavy sands extracted.[69] Private firms in this area were far less critical of the argument that justified state entrepreneurship on the basis of protecting national interests. Finally, civil construction remained safely the domain of private Brazilian firms.

In this section I concentrated on procurement as a strategy for favoring national private participation. I now turn to a review of other instruments that state agencies in Brazil could have relied on, to induce private sector participation in the nuclear program.

State Agencies and Sector-Specific Incentives

As I noted earlier, 1974 and 1975 were years of transition to import substitution in capital goods. This strategy implied not only pressures on state agencies to increase local purchases of capital goods, but also stricter import controls and greater availability of credit for investment.[70] In light of these general directives, how did state agencies perform in inducing local firms to provide inputs for the nuclear program?[71]

As the most powerful agency of the Ministry of Industry and Commerce in the execution of industrial policy, the Industrial Development Council (CDI) greatly influenced private investment during those years.[72] Although potentially a powerful instrument for the creation of "market reserve"—it approved credits and tariff breaks on imported equipment—the CDI granted such breaks indiscriminately until 1974.[73] The CDI's autonomy in industrial policy was quite limited, and its decisions required the consent of other agencies.[74] Thus the ministers of Finance, Industry and Commerce, Mines and Energy, and the Chief of the Planning Secretariat of the Presidency (SEPLAN) demanded preferential fiscal treatment for the nuclear program, similar to that accorded the oil and power sectors, whereby imported equipment was to be exempted from federal and state taxes.[75]

Nuclebrás requested the Council for Economic Development (CDE) to exempt it from the compulsory deposit on imports (Central Bank Resolution 354), from tax on industrialized products (IPI), and from imports tax, on all of Nuclebrás's purchases from German companies for power plants, for associated industrial facilities including NUCLEP, and for private Brazilian suppliers of components.[76] The CDE was another interministerial body, presided over by President Geisel, that coordinated activity among the Planning Secretariat and the ministries of Finance and of Industry and Commerce, among others. The president's Planning Secretariat functioned as CDE's Executive Secretariat.[77] The private sector was highly critical of this agency, which they regarded as beyond their influence.[78] CDE authorized Nuclebrás, in late 1977, with exemptions from the prohibition to import equipment with a "national similar."[79] A 1976 decree had restricted exemptions for machinery and equipment (even where a "national similar" was not available), although exemptions could be endorsed by presidential decree. In 1978, with President Geisel's approval, Nuclebrás and its subsidiaries were exempted from the compulsory deposit administered by CACEX and from import taxes on manufactured products. The total relaxation of import controls facilitated purchases of foreign equipment, machinery, blueprints, devices, and parts.[80]

The behavior of state agencies in the area of import controls thus reflected a pattern of overlooking extant legislation and suspending protectionist measures while granting preferential treatment to imports. Since 1974, resolutions by the CDI and CDE had emphasized support for national capital goods producers. CDI assistance to basic sectors increased from 5 percent (1973) to 18 percent (1977) of total investments (while assistance to automobile and consumer goods decreased from 5.5 percent to 2.6 percent). The participation of national industry in supplying equipment approved by CDI rose in that period from 35 percent to 68 percent.[81] However, it was not until 1979 that most fiscal exemptions for the importation of goods for priority projects by state companies were eliminated. The nuclear agreements, negotiated in 1974–75, preceded these steps, precluding prospective domestic suppliers from relying on import controls to secure a safe and predictable level of orders for the nuclear program. Effective orders for local firms were also delayed, in part, by uncertainty about who would assume responsibility for import taxes, including the 100 percent deposit on parts, equipment, and raw materials to be imported by national private firms. These firms requested either exemption from that tax, and from the Tax on Movement of Merchandise (ICM), or assumption of the taxes by the buyer, the state utility Furnas Centrais Elétricas.[82] The lateral segmentation in the program's implementation (the large number of agencies involved) increased the already high levels of "strategic uncertainty" confronting Brazilian firms capable of supplying inputs for nuclear power plants.

As for the provision of subsidized credit, the National Economic Development Bank (BNDE, later BNDES) had become the most important agent for strengthening Brazil's private sector (particularly capital goods) since the mid-1960s, and particularly during the 1970s. BNDE financed most infrastructural programs. Between 1974 and 1979, the share of BNDE funds captured by basic and capital goods grew from 48 percent to 69 percent, most of which went to private Brazilian firms.[83] BNDE's subsidiary FINAME (Special Agency for Industrial Finance) was created to finance, at fixed interest rates well below inflation, domestic purchases of capital goods. FINEP (Financial Agency for Studies and Projects), in turn, became a major financial agent for feasibility studies and project development.[84]

Nuclebrás had committed itself to obtain financial support for private firms from government institutions.[85] By mid-1977, however, Nuclebrás had not reached agreements with BNDE, FINAME, and the Bank of Brazil on the terms for financing domestically produced equipment. BNDE experts questioned the viability of NUCLEP (as a state firm), the incursion of Nuclebrás into areas where the private sector could have participated effectively, and the value of technology transfer agreements with German firms.[86] By 1978 no funds requested from FINEP, FINAME, or the National Housing Bank (BNH) had been released.[87] Eventually, the major private firms in the program benefited from some subsidized financing and fiscal incentives from FINEP and FINAME, to substitute for imports of materials and parts.[88] In some cases FINEP also financed purchase of equipment for, and training in, quality control. Not all firms, however, received financing for technological upgrading, and some relied on their own resources.[89] Overall, the level of support by FINEP for private producers in the nuclear sector appears to have been limited.[90]

Summing up, private Brazilian firms criticized the state's expansion into the heavy mechanical and engineering sectors, its procurement policies favoring foreign equipment and technology, the concomitant lack of import controls on foreign equipment and engineering services, and the deficient allocation of financing to encourage domestic purchases. National firms were not granted preferential treatment even when "national similars" were available. Where the price of locally produced components was higher than their imported equivalent (plus import and other incidental taxes, and technological upgrading), Nuclebrás opted for German equipment.[91] Nuclebrás could not even secure subsidized financing and fiscal incentives for private firms swiftly enough to minimize entrepreneurial uncertainty over nuclear investments. This pattern of state incentives (or lack thereof) to attract Brazilian private firms to the nuclear program, while protecting them from financial risk, stands in sharp contrast to the experience of industrialized countries in general and of Argentina in particular.

State Inducements and Private Entrepreneurial Response in Argentina

As argued earlier, private Argentine firms in the 1960s (when Argentina sought its first contracts for nuclear plants) were far less industrially mature than were their sectoral counterparts in Brazil in 1974–75 (when Brazil signed the nuclear agreements with KWU). Yet, for the Argentine Atomic Energy Commission (CNEA), national private participation was a cornerstone of the nuclear program, a policy hatched in isolation from secular trends in Argentina's industrial policy regarding capital goods.

Imports of capital goods were unrestricted from 1950 to the late 1960s. Since the late 1960s, Argentina encouraged its local firms slightly more strongly than Brazil, through tax incentives for the purchase of Argentine equipment.[92] By 1975, Argentine firms satisfied almost three-fourths of capital goods requirements. In the aftermath of the 1976 coup, imports climbed to one-half of all capital goods requirements (by 1980).[93] The argument that the nuclear program may have been compatible with broader sectoral trends, however, can hardly be sustained. CNEA drove a hard bargain with foreign suppliers on securing domestic inputs for the first nuclear plant (Atucha 1) during negotiations in the late 1960s, at a time when two-thirds of total imported machinery and equipment was exempt from duties, a good percentage of which had domestic equivalents.[94]

CNEA's motivations and capabilities are more easily understood when one recalls the low macropolitical consensus characteristic of postwar Argentina on the one hand, and CNEA's own lateral institutional autonomy on the other. These conditions allowed CNEA to mobilize private industrial clients and technical-scientific networks, while bringing its penchant for technical achievement to bear on the nuclear program. The longevity of CNEA's bureaucratic autonomy and the stability of the agency's principal—the Navy's leadership—ensured a remarkable continuity in CNEA's agenda of actively cajoling private firms to invest in nuclear-related production. CNEA sought to minimize "strategic uncertainty" through consistent and quite reliable procurement patterns. In spite of an initial lack of interest by formal producers' associations, CNEA officials coaxed a number of firms into the Argentine Association of Nuclear Technology. Since its creation in 1972, this association formalized a corporatist structure where CNEA officials, 54 private firms, and independent professionals worked out the fundamental principles of CNEA's protectionist design for a nuclear industry.

Exploiting its purchasing power, CNEA elaborated a system for maximizing the local content of prospective power plants and associated fuel-cycle facilities. Before the actual negotiations with a foreign supplier, CNEA officials constructed a "positive list" of mechanical, electrical, and electromechanical components that national firms were able to produce and that would result in their technological upgrading. CNEA then protected all items in this "positive list" (at times priced several times their freight-on-board equivalents) through a cost-plus fee system.[95] A second list, the "probable list," included components for which there was no absolute certainty that domestic production was feasible, but which would command special efforts to secure a national supplier. Rather than demanding a more financial percentage of local participation from the main contractor (Siemens and AECL), CNEA negotiators used a very coherent strategy of dissecting all components in a bid (unpackaging), with an eye to maximizing Argentina's inputs.

In promoting domestic supplies, CNEA had to rely mostly on its own resources; it was unable to secure incentives from the Industrial Secretariat of the Ministry of Industry and Commerce, or industrial promotion mechanisms such as those granted to the petrochemical, shipbuilding, and automobile sectors.[96] Thus, for instance, CNEA absorbed all initial costs for the domestic supply of a few heavy components (of the steam-generating nuclear system). The costs included the firm's upgrading and qualification, training (in-house and abroad), prototype development, materials acquisition, and technical assistance (and even some compensations) to the licenser.

Unlike Brazil, which created a state-owned producer of heavy components (NUCLEP), CNEA contracted a private Argentine firm to produce parts of the primary circuit (steam generators, moderator coolers, and pressurizers—see Appendix D) for Argentina's third nuclear power plant (Atucha 2). CNEA also organized a consortium of medium-sized firms in the area of instrumentation and control, and operated Services of Technical Assistance to Industry (SATI) to actively boost private firms in their technological development. Such services included preliminary comprehensive flows of information on systems characteristics, provision of testing, material selection, and services in welding, quality assurance, and technical assistance to production. In the fuel-cycle area, CNEA granted a private firm the right to exploit a low-grade uranium plant (Los Gigantes). Finally, CNEA trained industrial personnel in foundry, heat treatment, plastic deformation, modern physical metallurgy, and other areas.[97] Many of these activities were conducted prior to negotiations with foreign suppliers, in order to increase the agency's ability to bargain effectively for domestic inputs.

During the second half of the 1970s, CNEA began creating mixed-ownership enterprises (state/national private) in the wake of a general effort—with meek successes—by the Videla government to privatize state firms. First, CNEA established INVAP (Investigaciones Aplicadas) as a semi-public enterprise, a joint venture between the province of Río Negro and CNEA. INVAP's financing came from its own projects and feasibility studies; its R&D activities included the design of a 20 MW reactor.[98] Another joint venture, CONUAR, linking CNEA with a private concern (PECOM), began operating in 1983; with a majority of private capital, the firm operated the Nuclear Fuel Elements Factory (FECN). Created by CNEA to produce fuel for its power plants, the firm purchased equipment and technology from CNEA, which retained proprietary rights to the licenses and strategic control of the firm.[99] A third mixed enterprise, Nuclear Mendoza (CNEA, 51 percent and Mendoza province, 49 percent), was created in the uranium mining and processing sector. As a consortium of five national private firms, it designed and produced components for Argentina's second power plant, a reprocessing plant, and heavy-ion tandem accelerator. A fourth enterprise (70 percent private) was planned to manage the special alloys production facility. Finally, the joint venture in nuclear plant architect-engineering, ENACE (CNEA-KWU), was created as a temporary arrangement, with a view to eventually replace it with private Argentine engineering firms.[100]

CNEA's commitment to the principle of state subsidiarity—that is, intervening only when the private sector is unable or unwilling to do so—was atypical, particularly when compared to the industrial entities under the control of the Army's Fabricaciones Militares (DGFM), including steel and petrochemicals. I return to this apparent puzzle in Chapter 8. Yet, as the program evolved, and its associated private firms with it, even CNEA could not avoid criticism for expanding state control over potential private markets. CNEA was blamed for monopolizing activities which, although not specifically nuclear, had a bearing on some of the processes related to the fuel cycle, such as the extraction of zirconium, the production of special alloys, and some electronics components. Argentine industrialists, now more experienced than when first cajoled into nuclear products, argued that state control of these areas prevented their own expansion into nonnuclear products and processes. Similarly, private engineering consultants voiced their difficulties in penetrating the nuclear sector, owing to CNEA's attempt to control most technical aspects.

At this juncture, a discussion of the place of private firms associated with the nuclear sector within the broader spectrum of Argentina's producers' associations is in order. Argentine entrepreneurs were historically divided into two rival camps.[101] On the one hand, an umbrella group—the Asociación Coordinadora de Instituciones Empresarias Libres (ACIEL)—included the largest and most powerful commercial, agricultural, financial, and industrial interests. Associated with ACIEL was the Unión Industrial Argentina (UIA), which included the major transnational corporations and the largest national firms. This camp, often imputed with "hegemonic" aspirations among Argentine industrial sectors, had traditionally rejected state entrepreneurship and artificial state protection of uncompeti-

tive firms.[102] Both the powerful UIA and the Association of Metallurgic Entrepreneurs were lukewarm to CNEA's initiatives at the
outset.

On the other hand, the Confederación General Económica (CGE),
created by Perón in 1952, included more traditional industrial sectors, which opposed foreign capital and its domestic partners and favored protection and corporatist arrangements. In order to maximize
political impact, the CGE often relied on nationalist slogans against
"denationalization" and on calls for domestic control over economic
decision making and national sovereignty.[103] These "patriotic entrepreneurs" were CNEA's ideal partners: highly compatible with
politically appealing notions of military nationalism and regional
power aspirations in general, and with the Navy's subsidiarity approach in particular. This sector was more powerful in the Argentine provinces than in Buenos Aires, a fact that may well explain the
pattern of CNEA's joint ventures with provincial firms described
previously.

Negotiations with nuclear suppliers were often carried out under
an unstable tug-of-war between ACIEL and CGE.[104] The first power
plant (Atucha 1) was negotiated in 1967, under President General
Onganía, a leader of the Army faction that had defeated Navy forces
in the bloody encounter of 1962. In 1966 President Onganía ap-
pointed an economic leadership (Jorge Nestor Salimei to the Ministry of Economy and Francisco Rodolfo Aguilar to the Secretariat of
Industry and Commerce) that represented the small- and medium-
sized firms of national origin affiliated with the CGE.[105] By late 1966
pressures from liberal quarters led to Salimei's replacement with
Adalberto Krieger Vasena, who represented powerful entrepreneurial
groups that largely advocated greater foreign investment and technology and challenged state entrepreneurship. In August 1968 Onganía's regime shifted toward a more statist-corporatist, nationalist
model, bringing about the dismissals of liberal-minded Army com-
mander Julio Alsogaray and of the Navy and Air Force chiefs. Onganía's own ouster in 1970 led to a brief interlude of renewed national
corporatism under General Roberto Levingston, whose Minister of
Economy, Aldo Ferrer, championed support for national capital and
national private firms "to enter sectors of vanguard technology and
basic industry, which are those of greatest dynamism."[106] Ferrer
named prominent industrialists from the CGE to important positions (in the Central Bank, among others); not since 1955 was this

group so well represented within state institutions. A "Buy National" Law was adopted, requiring state agencies and enterprises to rely on domestic inputs and national technologies. However, dwindling budgets and the parochial microeconomic rationale of state agencies limited the effectiveness of that law.

The UIA was prominent in the opposition to Ferrer's policy of "Argentinization," with its statist, nationalist, and inflationary overtones. The coup that ousted General Levingston had strong support from the Navy and the Air Force. The new Army chief, General Alejandro Lanusse, assumed the presidency in 1971. His "Gran Acuerdo Nacional" (GAN) courted the CGE and corporatist arrangements and kept orthodox economic forces at arm's length. By 1972, however, the CGE—in alliance with Peronist forces—firmly opposed Lanusse and the GAN. Most other political forces backed the GAN in setting up elections for 1973. Perón himself returned to Argentina in November 1972 and won the elections in September 1973. This particularly schizophrenic period of Argentine history, with its contradictory pressures on state bureaucracies, provided the broad political background against which the second power plant (Embalse) was planned and designed in 1972–73 and ultimately negotiated in 1974.

The Videla coup of March 1976 initially brought about a more pure form of liberal orthodoxy in economic policy than its predecessors. For Economy Minister Martínez de Hoz, Krieger's policies had been no more than a more coherent, statist version of import-substitution-industrialization that had been attempted up to then.[107] Martínez de Hoz challenged the bloated and inefficient state sector and noncompetitive national private industry and dissolved (and outlawed) the CGE. Efforts at privatization and fiscal restraint threatened the nuclear and other programs, but the armed forces continued to protect their industrial fiefdoms from economic collapse.[108] Argentina's third nuclear plant (Atucha 2) was negotiated in 1979 under these adverse circumstances. The 1979 Nuclear Plan directed the construction of six new plants (beyond the first two) in a twenty-year span. The nuclear private sector, born a child of CNEA, was now politically stronger to pressure for continuity in state commitments to a national nuclear industry. Its firms were highly organized by the time a fourth plant was being considered in 1982. Grouped into eleven commissions, by specialty, they provided—upon CNEA's request—detailed assessments for a preliminary study of domestic supplies for that plant.[109]

The civilian administration of Raul Alfonsín assumed the presidency in December 1983, in the midst of unprecedented economic crisis and foreign debt. Alfonsín's tenure was characterized by heterodox policies, unilateral threats to suspend debt payments, and austerity programs. Reacting to CNEA's difficulties in reimbursing them for goods and services, nuclear firms intensified their activity in 1985, through the Argentine Association for Nuclear Technology as well as the Argentine Industrial Union.[110] This crisis resulted from the growing erosion of CNEA's lateral autonomy (and consequently, much of its financial independence) with the ascent of Alfonsín, who placed CNEA under civilian supervision for the first time in its history. The crumbling of CNEA's protective wand pushed those enterprises that were not diversified enough, and therefore relied heavily on nuclear supplies, to the brink. Although the group, as we have seen, did not rely on a solid power base within the Argentine Industrial Union, it enjoyed significant political visibility and became quite outspoken following the program's sharp slowdown in the early 1980s.[111] To some extent, the group's activities influenced the approval of a draft law by the Argentine Senate in 1985, which limited customs exemptions on imports of nuclear plant components and services.[112]

Peronist President Carlos S. Menem won the 1989 elections. As part of a comprehensive revolution aimed at extensive privatization, Menem's Planning Secretariat (under Vittorio Orsi) engineered the reorganization of CNEA and the privatization of some of its activities in 1991 and obtained Menem's approval to share control and oversight of CNEA investments (mostly in Atucha 2 and the industrial heavy water plant).[113] Raúl Boix Amat, director of the nuclear department at Techint (a firm nurtured by CNEA's subsidiarity principle), was one of Orsi's main advisers and helped design privatization schemes. These included, for instance, the potential transfer of Atucha 2 to Siemens for completion as a turnkey and the creation of a joint enterprise to transfer the three operating power plants to a private licensee, with CNEA handling only safety and repair issues. The uranium oxide production facility in the Córdoba Manufacturing Complex would be transferred to CONUAR (Argentine Nuclear Fuel Corporation), a firm now owned by the Pérez Companc group.[114]

Pérez Companc became Argentina's largest industrial concern by the early 1990s, diversifying its activities into oil and gas, construction, real estate, cement, steel, agriculture, hotels, and banking. It became a partner of multinational corporations and banks in the 1992 privatization of the state natural gas company (Gas del Estado). Pérez Companc and the expanded industrial group Techint both benefited from the privatization of the telephone system.[115] By the end of 1992 Argentina had privatized its state airline, telephone company, and electrical power and steel works, as well as a number of railroads and oil-drilling concessions, in one of the most active privatizing spurts in Latin America.[116] This general push toward privatization dismantled much of CNEA's traditional domain, even though—somewhat paradoxically—this was a logical outgrowth of the old policy of state subsidiarity.

This overview highlights two main points: (1) the low macropolitical consensus underpinning the sharp turns in Argentine economic and industrial policy, and (2) the ability of CNEA—with its lateral autonomy—to steer through this process, while endorsing a consistent sectoral blueprint. Such autonomy also allowed CNEA to subsidize private investments from its own resources and to minimize the high "strategic uncertainty" faced by most Argentine producers across industrial sectors.[117]

Shifting Involvements: State and Private Power

This chapter unravels the apparent paradox with which we started: that of Argentina's more consistent emphasis, relative to Brazil, on maximizing the role of its private firms, in spite of Brazil's better industrial endowments in most relevant areas. At the same time, the discussion raises a number of more generic questions regarding state–private sector interaction—and shifting involvements—in an industrializing context.[118] This last section concludes by presenting some of these issues, from the more specific to the more general.

First, state behavior in Brazil reflected the potential contradictions involved in maintaining macroeconomic and political stability on the one hand, and consolidating local private entrepreneurial political and economic power on the other.[119] The ability to sacrifice the short-term interests of a certain capitalist sector for the sake of maintaining the viability of the socioeconomic model as a whole revealed a measure of state autonomy. The attempt to benefit local capitalists was selective (in time and space) and subsidiary to the grand strategy of industrialization that the regime embraced as a source of legitimacy.[120] Private participation was restricted in the initial stages in order to minimize risks and costs, although national inputs were to increase with each new power plant (of the eight planned in the Brazilian-German agreements). The characteristics of the 1975 agreements were quite compatible with other projects involving capital goods during those years.[121] As Fabio Erber stated in an extensive study of this sector, "When the financing of such projects is linked to foreign suppliers' credits or to government-to-government bilateral trade agreements, the local producers are excluded."[122] The inflow of foreign loans since the mid-1970s, of which the nuclear program and other such large-scale projects captured a significant share, increased state leverage in defining private sector participation.

Second, the nuclear program exemplified the expansion of the state machinery that accompanied Brazil's greater integration into the world economy, far more than the denationalization of a strategic sector.[123] After all, in accordance with the 1975 agreements, a significant share of foreign components would have been progressively replaced with Brazilian inputs (public and private) in the long term. As in other industrial areas, the regime justified the creation of state enterprises (notably Nuclebrás's subsidiaries NUCLEP, NUCLEN, and NUCLAM) as a means of preventing denationalization.[124] This was compatible with a general tendency—not unique to newly industrialized countries—for states to intervene where both entrepreneurial risk and lump capital requirements are high, as is the case in the nuclear sector.

Third, notwithstanding the two preceding remarks, conflicts between bureaucratic and private forces in Brazil point to the maturation of an economically and politically significant entrepreneurial class, particularly since the 1970s. During the early 1970s, when the nuclear agreements were outlined and negotiated, central state institutions had privileged access to regulatory and sectoral agencies and were better able to neutralize pressures from the private sector.[125] Channels of entrepreneurial representation under President Geisel were too weak to allow formal private sector input, particularly during the decision-making stages.[126] ABDIB had very limited information on the terms of bargaining with German suppliers, and no access to the process, according to its president, Waldir Gianetti.[127] This was a "nonbargaining mutual adjustment" between state and private.[128] Yet the same agreements could have hardly been signed by the late 1970s, when the political strength of industrial entrepreneurs expanded in tandem with their productive capacity.[129] The erosion of macropolitical consensus since the mid-1970s broadened the internal elasticity of sectoral agencies in defining specific goals, while the access of interest groups to these agencies deepened. The increased involvement of over 50 private firms with CNEN's parallel program in the early 1980s provides tangible evidence of this shift.[130] By 1988, with the transfer of all power plants and the engineering subsidiary NUCLEN to the state utility Eletrobrás, Nuclebrás was dismantled, and NUCLEP and NUCLEMON were privatized. The fuel-elements plant was transferred to Uranium of Brazil, created as a subsidiary of Nuclear Industries of Brazil (INB), Nuclebrás leaner successor. The state companies NUSTEP and NUCLAM (Nuclebrás's joint ventures with German firms) were dissolved. In 1991 CNEN promoted the creation of a mixed (state-private) enterprise to develop and market nuclear technology (Rio National Laboratory).[131] Plans to fully privatize the construction and operation of Angra 2 (expected to become fully operational in 1997) gained momentum in 1993, whereas the construction of Angra 3 was discontinued altogether. Three remaining state enterprises—Uranium of Brazil, NUCLEMON, and NUCLEI, all subsidiaries of INB—were dismantled in early 1994. By 1995, Eletrobrás itself (and Angra 1 and 2 with it) were under consideration as targets of a large-scale privatization program launched under President Fernando H. Cardoso.

The mobilization of private entrepreneurs challenges the view that greater integration with international capital (which accelerated between 1964 and 1974) "saps" the local bourgeoisie politically.[132] Subsequent attempts to adjust their role in the nuclear program point to the capacity of national firms to influence the action of international (and state) capital, "even while being shaped by it."[133] Private sector criticism of KWU's control of Nuclebrás's subsidiaries hastened internal changes (within NUCLEN, for instance) in the direction of stronger technical and managerial control by Nuclebrás officials. Reduced KWU influence, in turn, helped improve the performance of "industrial promotion" departments within each subsidiary, designed to increase the participation of private Brazilian firms.[134] Limited as these adjustments may have been, they point to the national industrial bourgeoisie's bid for a leading role, particularly since the late 1970s. Brazilian industrialists "rode the coattails" of the nuclear program to broaden their own political sphere and embraced anti-statism with a vengeance once the ability of the state to subsidize private firms declined.[135]

Fourth, the Argentine case points to the difficulties involved in drawing simplistic inferences about the relationship between state and private interests from a study of the potential beneficiaries of state action. The private nuclear sector in Argentina during the nuclear program's early years was feeble and without political visibility in the industrial representational structures. Rather than attempting to influence the state, the private nuclear sector was created by it. Twenty years later, private nuclear entrepreneurs were ready to mobilize political support (and willing to endure some financial costs) to ensure the program's continuity. Some of these private firms grew to become among Argentina's largest industrial groups, well positioned to exploit the opportunities offered by the privatization revolution of the 1990s.

Fifth, suggesting that a variation in entrepreneurial behavior, in itself, may explain differences between the Argentine and Brazilian programs can prove misleading. Rather, as we have seen, opportunities and state incentives were stronger in Argentina, lowering the deterring effect of strategic uncertainty. CNEA's consistent commitment was able to offset, to some extent, the higher degree of market uncertainty (compared to Brazil) that had characterized Argentina's stop-go cycles since 1955. Postwar Argentina was never endowed with sustained entrepreneurial confidence in public policy. Instead, there was a deep mistrust—that CNEA succeeded in neutralizing—even of "developmentalist" military regimes.[136] This interpretation of entrepreneurial behavior carries with it two major implications. First, the initial reluctance by private firms should not be mistaken for an anti-industrial orientation: "a class interest in industrialization is not incompatible with a tendency to risk-averse investment behavior."[137] Thus, many an Argentine entrepreneur preferred cozy protective import-substitution strategies. Second, the role played by local bourgeoisie in large, relatively advanced industrializing countries was not much different from that of earlier developers, for whom industrialization required a no less interventionist state. More specifically, the behavior of nuclear entrepreneurs in Brazil and Argentina evokes the experience of their counterparts in the United States, who were not motivated merely by the expectation of technological benefits; rather, it was political and commercial incentives that helped mobilize the electrical equipment and utility firms into the country's civilian nuclear industry.[138]

There were, to be sure, important contrasts between the two countries' entrepreneurial structures and behavior patterns. Major sectoral producer associations—such as ABDIB and ABINEE—were active in Brazil, whereas in Argentina entrepreneurial access was channeled through a captive Argentine Association of Nuclear Technology, organized by CNEA, and not through broader sectoral organizations.[139] As we have seen, nuclear firms in Argentina were at the margins of major entrepreneurial associations. Thus, as Zysman suggests, "The political interests of each sector cannot be mechanically inferred; these interests depend on which types of firms control the industry associations and lobbies and on their own perception of their interests" (1977:293). Other firms in the sector may have different interests.[140]

Sixth, private firms in each country not only faced different types of incentives but also interacted with dissimilar state structures. On the one hand, a segmented arena of policy implementation (FURNAS, Eletrobrás, Nuclebrás, Ministry of Mines and Energy, BNDE) led to difficulties in coordinating financial and fiscal incentives, thereby increasing financial, economic, and strategic uncertainty for Brazilian entrepreneurs.[141] Whatever the efforts and motivations of the Nuclebrás officials responsible for "domestic industrial promotion," they were bound to be overwhelmed by the priorities of central economic agencies. Thus, the CDE (coordinating among the Planning Secretariat and the ministries of Finance and of Industry and Commerce) approved tariff exemptions for Nuclebrás, which, in turn, facilitated the import of foreign equipment and technology. President Geisel—former director of Petrobrás and an advocate of state entrepreneurship—headed CDE and wielded widespread influence on major economic agencies. On the other hand, Argentine firms could rely on solid signals from CNEA, whose lateral autonomy secured its financial wherewithal and capacity to bear costly inducement mechanisms. CNEA's awareness of structural barriers posed by central economic agencies forced it to rely on independent budgetary sources to finance the private sector. Argentine investors could thus commit their resources to *political* products while lowering their exposure to general market uncertainties.[142]

We can relate these conclusions to the argument postulated in Chapter 2 by characterizing graphically the relative efficiency of political signals under conditions of varying macropolitical consensus and bureaucratic autonomy, as in Figure 2. Signaling efficiency is highest in cases where bureaucratic autonomy is high, even if macro-

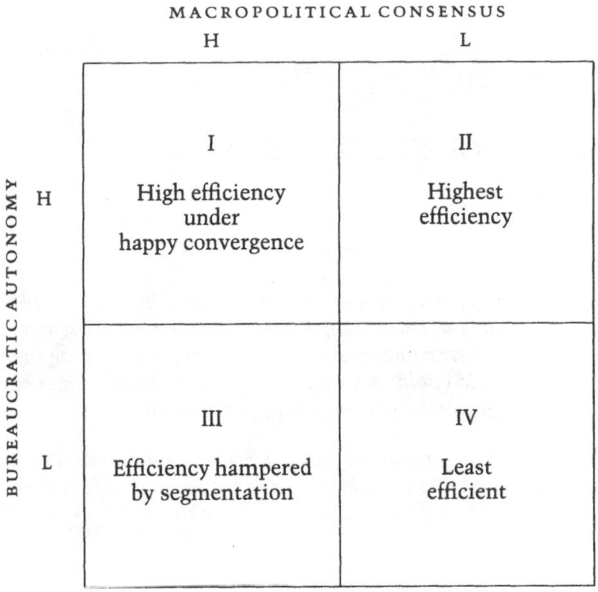

MACROPOLITICAL CONSENSUS

	H	L
H	I High efficiency under happy convergence	II Highest efficiency
L	III Efficiency hampered by segmentation	IV Least efficient

(left axis, vertical): BUREAUCRATIC AUTONOMY

FIG. 2. Efficiency of political signals to private entrepreneurs under conditions of high/low macropolitical consensus and bureaucratic autonomy.

political consensus is low (cell II), because the maverick agency is generally able to cushion private investors from uncertainty. Signals will be efficient if consensus is also high (cell I), provided a "happy convergence" reigns between central agencies (broad industrial patterns) and sectoral ones. Low bureaucratic autonomy and low macropolitical consensus provide the context for least efficient signaling (cell IV),[143] since the turn of bureaucratic struggles is often hard to predict. Low autonomy and high consensus (cell III) may allow more efficient signaling than in the previous case, although a high number of agencies may inhibit optimal efficiency.

Dissimilar state structures not only present private entrepreneurs with disparate sets of incentives and risks, but they also affect the international bargain with foreign suppliers of equipment and technology. It is to this international bargain that I now turn my attention.

CHAPTER 4

Bargaining in Technology: International Politics, Markets, and Regimes

There are some grounds ... for believing that the physical and economic factors (i.e., foreign exchange difficulties) which otherwise result in an extremely uneven distribution of energy supplies among the nations of the world could be greatly modified by the advent of commercial atomic power. (Schurr and Marschak, 1950)

The nuclear program accounted for over five percent of Brazil's total foreign debt ($130 billion), and for over seven percent of Argentina's public foreign debt ($70 billion), in the mid-1980s.[1]

As Putnam argues, "any testable theory of international negotiation must be rooted in a theory of domestic politics" (1988:442). The preceding chapters examined the origin of Argentine and Brazilian preferences for different international strategies and clarified the nature of domestic groups and institutions required to ratify an agreement with foreign suppliers of nuclear technology. In this chapter I begin by analyzing the specific impact of macropolitical consensus and lateral autonomy on the bargaining process. I then examine the particular characteristics of bargaining in the context of technology transfer. On the demand (technology-recipient) side, I focus on the implications of each state's strategy for the selection of technology transfer mechanisms. On the supply (technology-provider) side, I probe the international market context and the financial and political environment within which negotiations took place. Subsequently I evaluate the impact of that environment on the pricing of technology and on the acceptance of "special (restrictive) clauses." Understanding the terms under which technology transfer takes place is an essential component in assessing the way technology transfer affects the development of a recipient's national technological capabilities; this assessment is undertaken in Part III.

Macropolitical Consensus and Lateral Autonomy: Implications for Bargaining

Levels of bureaucratic autonomy and macropolitical consensus can influence the process of bargaining with technology suppliers through their impact on the size of domestic win-sets, on the risks of involuntary defection, and on the credibility of commitments and reduction of uncertainty.[2] For example, low consensus and high levels of autonomy (cell II in Figure 3) can narrow the size of the win-set to the institutional preferences of the sectoral agency. A small domestic win-set, in turn, can be a bargaining advantage and can increase the negotiators' leverage over the distribution of benefits from the international bargain. High levels of lateral autonomy, even in the midst of low consensus, cancel out the risk of involuntary defection, the small size of the win-set notwithstanding.

For over 30 years Argentina's domestic win-set in nuclear negotiations was restricted to CNEA's choices. The need for CNEA's (and the Navy's) single-handed endorsement of any nuclear agreement was clear to suppliers. The agency's autonomy in decision making and implementation—unencumbered by political and bureaucratic interference—strengthened the credibility of its commitment. Argentina's Ministry of Foreign Relations followed CNEA's instructions to the letter; the fact that this ministry was also in the hands of the Navy explains the smooth, unchallenged management of international bargaining on nuclear matters by CNEA officials.[3] CNEA's monopolistic control over the nuclear industry was far from a state secret; it was often invoked by the Navy for domestic purposes and openly wielded in negotiations with technology suppliers. Such control enabled CNEA to bargain forcefully and effectively for its preferences: maximum local content and national control of the technology transfer process. CNEA was able, for instance, to stave off technology suppliers' attempts to maximize foreign inputs through suppliers' credits, which often tie financing to the purchase of equipment. It was also able to resist attempts by suppliers to apply restrictive clauses, such as export restrictions, restrictions on the use of know-how once the contract was over, appropriation of improvements, and control over quality assurance. Foreign control over quality assurance confers to suppliers the ability to act as gatekeepers and to exclude local firms on technical grounds. CNEA even compelled

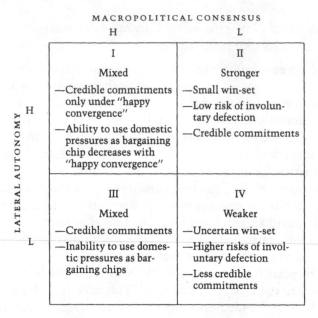

FIG. 3. The effects of macropolitical consensus and lateral autonomy on bargaining position.

suppliers to extend the quality guarantees accompanying their own equipment to equipment of Argentine origin.

Bargaining advantages may dissipate when both consensus and autonomy are low (cell IV in Figure 3). This combination increases uncertainty regarding the contours of the win-set, raises the risk of involuntary defection, and weakens the credibility of commitment. Under such conditions, unstable demands create a nightmare for the foreign partner (in our case, the supplier of technology), who is now forced to reassess continually the interplay among political and bureaucratic forces within the recipient state. The bargaining process under these conditions is likely to become protracted and unpredictable. As Putnam (1988) suggests, at a particular level of uncertainty, the unpredictability of the boundaries of the domestic win-set can lead suppliers to require additional assurances (side-payments) that ratification will take place. Such an environment may explain, for instance, Mexico's weakened bargaining position in negotiating agreements in the pharmaceutical industry.[4]

The independent effects of high macropolitical consensus (left column in Figure 3) seem to be mixed. On the one hand, high consensus within a ruling coalition (in actor A) may dissipate fears of involuntary defection in the opponent (actor B). There is greater certainty that A's coalition "can deliver," and this strengthens its bargaining position.[5] Manufacturing investors in Singapore, for instance, face little uncertainty about the nature and extent of the country's consensus over industrial policy. On the other hand, because high consensus improves the chances of easy ratification, A's negotiators are less able to use domestic pressures as a bargaining chip to obtain growing concessions from B. The effects of strong consensus, therefore, may be better gauged in conjunction with levels of bureaucratic autonomy. Where consensus is strong, high bureaucratic autonomy increases the risks of involuntary defection, makes the environment less predictable, and weakens the advantages of consensus. This is likely to be the case where there is little happy convergence between the substance of the consensus and the agency's institutional preferences. However, even at relatively high levels of macropolitical consensus, bureaucratic segmentation (or low sectoral autonomy) may dilute the clarity of the boundaries of the win-set. The risks of involuntary defection grow—weakening the bargaining position—but this effect is offset by the rising ability to use domestic dissent as a bargaining chip. Some of the difficulties imposed by a segmented institutional decision making—despite a fairly consensual development strategy—are evident from India's bargaining with foreign suppliers in the computer sector and from Mexico's bargaining in the automobile industry.[6]

Brazil's more or less consensual industrial strategy in the late 1960s and early 1970s arguably strengthened its credibility vis-à-vis foreign suppliers. This is the background against which Brazil negotiated the so-called deal of the century—estimated at about $14 billion in 1975—with KWU and its associated financial and industrial partners.[7] At the same time, German negotiators appeared to have discounted the possibility of domestic Brazilian opposition to these agreements. In particular, they assumed (on the basis of exploratory contacts with state officials) that the boundaries of Brazil's win-set were quite flexible.[8] Indeed, these boundaries were flexible, given the perceived high costs of "no agreement" for the fate of the ruling coalition's political-economic model.[9] Nevertheless, extensive Brazilian concessions to German imports, the creation of foreign-controlled state-foreign ventures, and the acceptance of restrictive clauses in technology transfer and of an unproven uranium-

enrichment technology had placed the agreement beyond the consent of many private Brazilian industrialists and national technical
networks.[10] Thus, although Brazil's consensus was beginning to shat-
ter by 1975, Brazilian negotiators failed to use domestic pressures
to exact concessions that would have satisfied important national
constituencies.[11] Efforts by some Nuclebrás officials to heed those
groups' demands were suppressed by a highly segmented decision-
making context, which echoed the general directives of central state
agencies.[12] These guidelines implied, as we have seen, a built-in preference for foreign capital, equipment, and technology, capable of establishing a nuclear industry within the shortest possible lead time.[13]
Consequently, even a coherent opposition on the part of Nuclebrás—
one not evident at the time—was likely to have been overwhelmed
by the combined weight of central economic agencies.[14]

  If negotiators are not always clear about the boundaries of their
own domestic win-set (as appears to have been the case in Brazil),
they can be even more misinformed about the complexities of domestic games on the other side.[15] High consensus within actor A may
make it easier for negotiators from actor B to assess the domestic
win-set of A but does not preclude miscalculations. For example,
German negotiators (like Brazilian officials) failed to anticipate how
a loosening macropolitical consensus was rapidly changing Brazil's
win-set. They thus pressed for levels of German control, equipment
supplies, and technological inputs that went beyond Brazil's effective—rather than presumed—win-set. What German negotiators did
not miscalculate was the relative weight of Brazil's military establishment within the ruling coalition. They understood at the outset
that the "deal of the century" would require the transfer of complete
nuclear fuel-cycle technology (including sensitive capabilities such
as uranium enrichment and spent-fuel reprocessing).[16] This "synergistic issue linkage,"[17] in turn, compelled the Germans to broaden
their own win-set in such a way that it would include the transfer of
such technologies. That step required German government support
for a transaction that challenged U.S. nuclear exports policy. This
challenge may well be considered one of (now united) Germany's
early steps in the direction of growing political assertiveness in
world politics. The Brazilian deal was also a precursor to German
firms' extensive practice of exporting sensitive technologies—nuclear and otherwise—which, in the 1980s, included recipients like
Iraq and Iran.

German willingness to provide Brazil with complete nuclear fuel-cycle technology points to the other side of this two-level game, or to the ways in which Brazilian and Argentine negotiators used the bargaining process to rally support (or at least minimize challenges) from domestic groups. The Brazilian team was well aware that a military regime had been orchestrating Brazil's consensus after all, even if negotiations on all industrial programs—including nuclear—were primarily in the hands of high-level civilian technocrats. In Argentina, CNEA wielded the foreign partners' commitment to an early participation of national industrial, scientific, and technological resources as a means of coalescing, at the outset, a wide front of support for the nuclear program—and for CNEA's institutional trajectory.

Given this overview of the general contours of the bargaining context, I now turn to a more detailed analysis of the technological dimension of bargaining, from the point of view of the recipient.

Technology Transfer: The Demand Side

State enterprises have been both major purchasers of custom-built capital goods and engineering services and suppliers of essential inputs to the rest of the economy, particularly in industrializing countries. As such, these enterprises have used procurement as an instrument of technology policy, "to the extent that a deliberate action is taken to steer the externalities produced by state purchases towards fixed aims."[18] Major interrelated goals of state intervention in technological development include offsetting deficiencies in the market for technology, protecting the interests of technology recipients, and strengthening national technical capabilities. The instruments available to advance these goals include encouraging demand for local technology, increasing the capacity to absorb technology, regulating technology imports, and guiding the production of local technology.[19]

The "whitening of black boxes"—the attempt to disaggregate, as much as possible, the technological components of a project in order to maximize national technical inputs—has been considered a major instrument of regulating technology imports.[20] This strategy is based on the premise that the selection of a particular technological package implies a series of concomitant decisions over systems, components, and materials—modules—all of which are interrelated, but which can be examined independently.[21] The partial independence of this series of technological decisions enables a better gauging of those areas in which domestic firms can perform effectively. Thus, unpackaging or unbundling the investment is viewed as an essential instrument to maximize localized learning and national participation.[22] In the nuclear sector, for instance, this strategy implies inviting international bids for specific areas (the turbogenerator portion or the nuclear steam supply system) only after completing domestic bids inducing local firms to supply components.[23] Where domestic capacity is inadequate, technological growth can be achieved by ensuring maximum absorption and diffusion of the imported technology—through, for instance, joint foreign-local ventures—with the recipient gaining information on basic designs and effective technical assistance.

The level of packaging is considered to be higher as the technology supplier provides more components of the peripheral (i.e., non-core) technology. Only basic engineering, training, start-up, and troubleshooting services are considered core technology.[24] Prior investments in learning about the technology in question help minimize the ability of foreign suppliers to capture the provision of peripheral items. Although some consider direct purchases of embodied technology (such as machinery and equipment) to be black boxes, others regard embodied technology as an inexpensive way of learning and as a model for reverse engineering.[25]

These considerations regarding the effective use of national technological and productive resources were central to CNEA's bargaining postures and served CNEA's domestic political clientele well. CNEA's preferences constituted the narrow domestic win-set guiding Argentina's bargaining for consistent unpackaging of foreign technology and for maximizing national R&D efforts and learning-by-doing. Argentina's training programs and early in-house performance of feasibility, marketing, and other preliminary studies aimed at achieving decreasing levels of packaging. CNEA's negotiators regarded technical assistance by suppliers in disaggregating the package as a key consideration in evaluating bids.[26] The fact that the same technical group conducted the feasibility study, the evaluation of bids, and the contractual agreements, as well as supervised construction, resulted in greater coherence and continuity of purpose.[27] Although Argentina's first two reactors (Atucha 1 and Embalse) involved turnkey contracts, CNEA insisted on an effective role as prime contractor, without limiting the supplier's overall responsibility for the project.[28]

Without CNEA's lateral autonomy, Argentina's win-set would have been likely to contain the option advanced by Argentine electrical utilities, that is, maximum imported equipment and light water reactors.[29] CNEA's autonomy sheltered it from the country's cyclical political turmoil, and its continuity in leadership prevented the conceptual shifts that characterized Brazil's nuclear policy since the 1950s. Argentina purchased its first nuclear power plant (Atucha 1) in 1968, under Onganía's military rule; it selected Siemens's offer of natural uranium technology despite a lower bid by Westinghouse. In 1973, under a civilian (Peronist) regime, CNEA negotiated a contract for a second power reactor (Embalse) with a new supplier—Canada's AECL-IT—preserving the natural uranium line. During the military administration of President Videla, in 1979, CNEA contracted with KWU (a Siemens subsidiary) for its third plant (Atucha 2), also a heavy water reactor, and with the Swiss firm Sulzer for a heavy water production plant, to ensure independence in fuel supply for all power plants.

In contrast, Brazil's win-set compelled (in the mind of its negotiators) an agreement responsive to the challenges imposed by the post-1973 oil debacle, a crisis that impaired the macropolitical model of industrialization. The emphasis on, and success in, reducing foreign suppliers' leverage in the provision and subcontracting of peripheral items was comparatively weaker. The technological externalities of domestic production were, in Brazil's case, secondary to the overall preference for integrated timetables and reliable technology. This preference guided negotiators, who represented not merely the sectoral agency Nuclebrás, but a highly segmented decision-making context. Critics had assailed CNEN since the late 1960s for failing to attract competent expertise in the technical and commercial areas of nuclear energy.[30] In any case, a few of these nuclear experts had an effective voice in this process.[31]

That levels of domestic content and technology transfer considerations played a subsidiary role, at best, is evident from an analysis of the five joint ventures between Nuclebrás subsidiaries and German firms, created by the agreements with KWU. The five were established in different areas of power reactor and fuel-cycle activities: NUCLEN (engineering), NUCLEP (heavy components), NUCLEI and NUSTEP (isotopic enrichment), and NUCLAM (mining). Although

KWU was a minority partner in all but one (NUSTEP), it exerted great influence over Nuclebrás by virtue of its defined participation in the decision-making process, in the choice of local partners, and in quality assurance qualification procedures. For instance, the shareholders' agreement for NUCLEN (75 percent Nuclebrás, 25 percent KWU) left KWU with substantial control over managerial and structural matters.[32] First, the German partner could nominate the technical and commercial directors, an arrangement severely limiting the power of Brazilian directors.[33] Most top management positions were held by KWU officials. Second, some items required unanimous (or super-majority) decisions, providing KWU with a virtual veto power and the prerogative—through its control of NUCLEN's administrative council and technical committee—of overriding decisions by the directorship. Finally, KWU was recognized as the sole provider of technology, and any export of services by NUCLEN, or any cooperative agreement with similar enterprises abroad, required KWU's approval. In turn, the penalties assigned to KWU for errors in conception were minimal.

Joint ventures can have more limited positive effects on local technological development if considerable control is retained by the foreign partner.[34] The central question, however, is whether the constraints imposed by joint ventures necessarily impair localized learning processes. Brazilian officials at NUCLEN claimed that the very establishment of a joint venture was geared to ensure technology transfer, compelling the foreign company to provide its name and capital and thus guaranteeing its stake in the firm's success and the quality of the product.[35] If anything, joint ventures were a favorite instrument of the macropolitical consensus reigning throughout the 1960s and early 1970s. Joint ventures sprang up in petrochemical and big mining projects and were often regarded as advantageous by both suppliers and recipients.

The outcome of the bargaining process, of course, was determined not merely by the recipient's respective technological preferences and instruments, as described in this section. While the global nonproliferation regime had a somewhat marginal impact on negotiations about domestic content issues, conditions in the nuclear technology and capital markets played a much more significant—but perhaps unexpected—role. It is to these international aspects that I turn now.

Technology Transfer: The Supply Side

INTERNATIONAL POLITICAL CONSTRAINTS

In what ways would international power politics and the nonproliferation regime have affected the bargaining positions of nuclear technology recipients? This question is relevant because international nuclear technology transfers have never been the result of the free operation of market forces. On the demand side, state intervention and mercantilistic objectives—rather than free markets—have been the rule, both for old-timers and newcomers to nuclear technology.[36] On the supply side, the diffusion of nuclear technology was subject to controls, initially through the unilateral policies of pioneer countries and later through multilateral regimes.[37]

The United States had maintained a policy of technological denial since 1946, following the rejection of the Baruch Plan, and until Eisenhower's Atoms for Peace program in 1953. The 1953–74 period, in turn, has been characterized as one of international promotion of civilian nuclear energy which, since 1957, came under the aegis of the International Atomic Energy Agency (IAEA).[38] In exchange for nuclear technology and materials, countries agreed to accept internationally applied safeguards. The Nonproliferation Treaty (NPT) of 1968 established a new set of guidelines for nuclear trade, securing access to nuclear technology for NPT signatories, by diluting—or nullifying—their sovereign rights to reject intrusive international inspections.[39] Much less clear was what the relationship between technology suppliers and nonsignatory recipients would be. In practice, for countries outside the treaty, only materials and facilities related to the particular transaction supervised by the IAEA were placed under safeguards. In other words, other indigenous (or copied) nuclear facilities remained outside international supervision, eschewing the concept of full-scope safeguards.[40] Not until 1977 did the London Suppliers Group agree on a more detailed code of conduct for nuclear technology transfers, by declaring a commitment to exercise restraint in the transfer of sensitive technologies (such as uranium enrichment and reprocessing of fuel waste) that might lead to the acquisition of weapons-grade material.[41] Suppliers could veto the ability of recipients to retransfer exported items and derivatives without prior consent. But even these new trading constraints did not include the imposition of full-scope safeguards on recipients as a

condition for trade. Countries like India, Brazil, and Argentina—
none of whom signed the NPT—have regarded this regime as an attempt by advanced countries to preserve their technological and
commercial monopoly through cartel-like arrangements, such as the
London Suppliers Group.

Clearly, this brief overview of the global nonproliferation regime
suggests that no serious analysis of emerging nuclear industries can
ignore the importance of international political considerations in the
transfer of this technology. The critical question for the purpose of
this study, however, is the extent to which understanding these conditions helps explain variations in Brazilian and Argentine bargaining performance and the resulting mutual adjustments among state,
foreign, and national private sectors. In reality, the injunctions of this
international security regime had only a marginal impact on the industrial and technical characteristics of nuclear programs of newcomers in general and of Brazil and Argentina in particular.[42]

First, similar international conditions (prior to 1975) allowed the
concomitant definition of different domestic strategies in Brazil and
Argentina, regarding both the allocation of power plant inputs and
the choice of technology. Contrasting decisions between Argentina's
first two plants (Atucha 1 and Embalse) and Brazil's first (Angra 1)
and eight (planned) consecutive plants were taken during the late
1960s and 1970s, under the same international constraints outlined
earlier. Greater indigenization ratios or domestic content requirements (particularly of conventional components) than those stipu-
lated in Brazil's agreements with KWU were perfectly compatible—
as the Argentine case reveals—with the international political regime reigning at the time.

Second, mutations in the international regime over time did not
lead to fundamental changes in those initial strategies. International
legal restrictions grew with the approval of effective procedures and
sanctions by the London Suppliers Group (1977–78), with the establishment of the International Nuclear Fuel Cycle Evaluation (INFCE)
in 1977, and with the passing of the Nuclear Nonproliferation Act in
the United States (1978); these restrictions made nuclear trade conditional on the acceptance of full-scope safeguards by the recipient.[43]
Yet the more stringent international conditions of the late 1970s did
not change the basic industrial and technological parameters of the
Argentine program or, for that matter, did not prevent the transfer of
a Swiss heavy water plant to Argentina in 1979–80, filling the last

technical gap in a comprehensive program defined twenty years earlier.[44] Neither did these particular international conditions transform the fundamental nature of the 1975 arrangements for the nuclear sector regarding state and private, and local and foreign, entrepreneurship or technical industrial paths. Had international political constraints—such as suppliers' reluctance to transfer enrichment technologies in 1975–76—played an overwhelming role, they would have perhaps altered Brazil's choice of technology in favor of heavy water cycles, which were subject to fewer restrictions in international nuclear trade at the time.[45]

This is far from arguing that the international nonproliferation regime had no impact whatsoever on nuclear technology transfers. In the first place, no transfer could have taken place legally without attached IAEA safeguards. More specifically, the policies of the Carter administration placed constraints on the range of potential suppliers. Brazil's attempts to acquire four nuclear plants and fuel-cycle technology from U.S. firms in 1974 failed when the U.S. administration vetoed the transfer of enrichment technology.[46] Later that year negotiations with France over enrichment facilities collapsed over French insistence on a turnkey plant, which Brazilian negotiators rejected. These pressures and conditions narrowed the choice of technology suppliers to West German firms, leaving them with a somewhat monopolistic position in the provision of eight light water reactors and complete fuel-cycle (including enrichment) technology.[47] International pressures may have also accounted for Brazil's acceptance of an unproven commercial enrichment technology (the German jet-nozzle procedure) instead of another technology (centrifuge) that would have required Dutch and British consent. However, these pressures did not prevent the transfer to Brazil of enrichment technology per se and may have contributed in accelerating Brazilian efforts—in the late 1970s and early 1980s—to seek independence in sensitive technologies through its so-called parallel (secret) program. In the end, efforts by Presidents Ford and Carter to cancel or revise the 1975 Brazilian-German agreement failed.[48]

The ability of both Brazil and Argentina to undermine German commitment to the nonproliferation regime was no small success, especially at a time when Germany's foreign policy was defined by the parameters of Pax Americana. Understanding this success (and much of the weakness of the international political regime that I have analyzed), as well as some of the bargaining advantages that nuclear technology recipients enjoyed throughout those years, requires an examination of the structure of the international nuclear market.

THE INTERNATIONAL NUCLEAR MARKET

The role of multinational corporations (MNCs) is as much at the heart of nuclear technology transfers as it is in other industrial sectors. Once stripped of the kinds of political constraints just discussed, the laws of demand and supply—even for reactor cores—function as with any other commodity. In fact, a great number of electromechanical components and services (about 80 percent of a nuclear plant) are of a conventional (nonnuclear) nature, and their study evokes many of the generic debates regarding capital goods and engineering markets.[49] In this respect, some of the general observations about bargaining between recipients and suppliers in the international political economy literature should be directly applicable to a study of bargaining for nuclear technology. If so, what would some of these general observations suggest to us about the differences between Brazil and Argentina in bargaining behavior and outcomes?

Central to the literature on technology transfer is the concern with relative bargaining power between recipient and supplier.[50] A recipient's ability to bargain effectively is expected to decrease under two conditions: when the bargain is over highly sophisticated technology[51] and when suppliers are highly concentrated.[52] The nuclear sector combined both characteristics (high technology and high concentration), particularly in the mid-1970s, placing—at least according to this twofold proposition—severe constraints on recipients.

The international nuclear market includes three main branches: uranium mining, the fuel cycle, and reactor development and manufacture, including associated thermomechanical and electrical equipment and components.[53] Producers of capital goods for electric power generation, rather than the uranium mining and chemical firms in the nuclear fuel cycle, were behind the expansion of nuclear power in the 1960s. This analysis of the negotiating process focuses to a considerable extent, therefore, on reactor development and manufacture. The robust and concentrated nature of the international power plant market, which subsumes nuclear electricity generation, preceded the arrival of nuclear energy. High concentration may have helped the industry transfer its unofficial oligopolistic arrangements at home to the world arena, in the form of price-fixing, quotas and allocations of bids, compensations (by winners of a bid to unsuccessful tenderers), and pooling arrangements.[54] High costs and risks, and uncertainties over market conditions, strengthened concentration and barriers to new entrants.[55] Main supplier states (except the United States) backed the technologically dynamic, self-regulating industry and tolerated the international cartel. These apparent bargaining advantages by nuclear technology suppliers, however, were eroded by an array of countervailing considerations.

In the first place, while U.S. and British firms dominated nuclear supplies in the immediate post-war period (the U.S. held 90 percent of the market in the 1960s), by the mid-1970s West German, French, and Canadian firms were effectively competing with U.S. firms, both in reactor and fuel-cycle technology.[56] Expanded production led to fierce competition among the few electrotechnical producers and eventually to excess capacity. The slow growth of domestic markets fueled the industry's attempt to persuade utilities of the sector's international viablility by stressing the potential for expanding political and economic relations with developing countries and increasing national prestige. Competition intensified because of the small number of transactions, their zero-sum nature, and the large financial stakes involved.[57] Despite euphoric projections in the 1960s and 1970s, demand grew far below the expected rate. This host of political and economic considerations exacerbated a competitive nuclear supply structure.

Within this buyers' market, advanced industrializing countries had a particular appeal as the last reservoirs of commercial opportunity—a condition that strengthened their bargaining position even further.[58] To improve their own relative positions, suppliers were willing to relax safeguards and provide "sweeteners" (in the form of sensitive technologies), to offer generous credit terms, and even to barter reactors for other commodities.[59] This environment broadened the opportunities of rapidly industrializing countries like Brazil and Argentina to cast the transfer of nuclear technology and equipment in accordance with their domestic industrial and technical priorities. For Argentina, supply-push factors in the 1970s facilitated the achievement of objectives defined earlier, in the 1950s. Conditions preceding this heightened global export market, however, did not prevent Argentina's CNEA from successfully pursuing those same objectives in the 1960s during negotiations for its first power plant.

Brazil's large-scale industrial program was defined at the height of the permissive market conditions, in the mid-1970s, when nuclear exports were critical to the industry's survival in most European countries. KWU—the major German supplier at the time—was particularly vulnerable, being the firm among its peers in Germany most dependent on foreign markets.[60] With its main fixed assets in turbine generator manufacture, its export platform incorporated large nuclear components manufactured by associated firms, including Steag, Kewa, Interatom, UG, GHH, Uhde, GS, and the Austrian firm Voest-Alpine, while electronic equipment was the reserve of KWU's parent company, Siemens. KWU accounted for about 25 percent of the costs of a single plant built in Germany, while over seven hundred other German companies accounted for the remaining 75 percent.[61] As Walker and Lonnroth (1983) suggest, subcontracting was an important means to minimize heavy fixed investment and to increase KWU's attractiveness as a partner abroad. Securing export shares for German firms was a core priority for German negotiators; their eagerness to win Brazilian contracts was backed by German credit and lending agencies and by the German government. Brazil was a major export market and Germany's most important trading partner in the industrializing world.[62] Clearly, the costs of no-agreement were quite high for the technology supplier.

On the one hand, Brazilian negotiators arguably took advantage of at least some of these conditions by insisting on the transfer of complete full-cycle technology and by accepting abundant German credit. Moreover, the willingness of KWU and associates to establish joint ventures in power plant engineering, heavy component fabrication, and the nuclear fuel cycle reveals a measure of adaptation to Brazil's objectives.[63] The attainment of high levels of domestic content is not a reliable standard for evaluating Brazil's bargaining success or failure. As we know from the preceding chapters, this criterion was not central to Brazilian negotiators in the first place, a fact that helps explain why the opportunities for national entrepreneurial participation were not exploited. Domestic institutional considerations also explain why demanding complete full-cycle technology—a major component of the domestic win-set—was an absolute requirement.

On the other hand, Brazil failed to realize the opportunities granted it by international nuclear market conditions when it acquiesced to: technical committees dominated exclusively by Germans (with veto power regarding the provision of equipment and services);

an economically unproven enrichment technology; the suppliers' limited assumption of risks; and overpriced imported equipment and training.

Financing arrangements can play an important role in shaping an industrial sector, and the extent to which they shed light on Brazilian and Argentine bargaining behavior is worth considering here.[64] Domestic constraints on financing for the nuclear program were common to both countries, although Brazil was particularly affected by the 1973 oil shock and its aftermath. Both countries relied on indebted industrialization as a generic strategy, particularly in the latter part of the 1970s, when Euromarkets were flooded with petrodollars.[65] Yet Argentine and Brazilian negotiators differed in their relative receptivity to suppliers' credits because such credits often displace local firms by tying financing to the purchase of equipment. Since integrating national firms was a primary consideration in Argentina, CNEA resisted the supply-push effect of suppliers' credits. For instance, starting with their first plant, CNEA both compelled Siemens to finance the production of components produced locally and assumed direct responsibility over differences in price. CNEA generally favored independent financing and sought to cover over 50 percent of the costs of its third plant from such sources.

Concerned with the rapid development of a nuclear industry to help stem a fizzling economic miracle, Brazilian negotiators were more amenable to suppliers' credits, which would presumably enable Brazil to undertake a large-scale industrial program without unnecessary delays. They thus chose to forgo a search for independent financing, despite Brazil's status as a favorite of international banking during those years. By 1976 estimates for the construction of the first two plants under the agreement with KWU required a total of $2.6 billion, one-third of which had to be sought domestically.[66] About two-thirds ($1.7 billion), to be backed by national treasury guarantees, came from the state Deutsche Bundesrepublik Bank (45 percent), a consortium of 30 private German banks headed by Dresdner (45 percent), and branches of German banks in Europe (10 percent).[67] The first financial agreement (with Kreditanstalt) included a clause forcing Brazil to spend at least 90 percent of the funds in Germany.[68]

The high volume and long maturity period required by these loans turned the Brazilian-German nuclear agreements into landmark financial transactions in the mid-1970s. A major borrower was the electrical utility Furnas Centrais Elétricas, the actual client purchasing nuclear reactor components and services, backed by guarantees from its holding firm Eletrobrás. The pressures on Nuclebrás's budget brought about by Planning Minister Delfim Netto's policies increased the need to turn to foreign financing, including suppliers' credits. At home, delays and reluctance to disburse approved financial credits were not unique to the nuclear sector.[69] Above all, the financial aspects emphasized the highly segmented decision-making and implementation context of Brazil's nuclear program.

This overview of international market and financial considerations underlying the bargaining process in both cases suggests that recipients were not particularly incapacitated to counter the oligopolistic practices of nuclear technology suppliers. In fact, both recipients succeeded in softening political commitments of the West German state vis-à-vis the nonproliferation regime. Moreover, international financial arrangements—as in many other industrial areas—served state objectives, whether those embedded in Brazil's macropolitical consensus or in the preferences of Argentina's highly autonomous state agency. Indebted industrialization had no fixed effects and could have been used to expand state entrepreneurship or private ownership.[70] However, a more detailed analysis of the restrictive conditions of technology transfer accompanying the Brazilian-German agreement sheds light on the sinuous ways in which an aggressive supplier can offset a recipient's advantages.

Evaluating Restrictive Clauses and Costs

The imposition of restrictive clauses and overpriced technology and equipment is often held as primary evidence of the vulnerability of technology recipients to predatory practices by MNCs. The unpackaging and screening of foreign technology not only benefits local technological resources; it is also a mechanism that helps evaluate cost structures more efficiently and minimize restrictive clauses (such as export prohibition, the appropriation of improvements, restrictions on use of know-how once a contract is over, and so on).

CNEA involved an effective team—designed to limit the use (and abuse) of standard commercial practices by technology suppliers—at the bargaining table.[71] The team sought turnkey contracts while insisting on an effective role as prime contractor, without limiting the supplier's overall responsibility for the project. In contrast, Brazilian negotiators yielded to a variety of quality control clauses, exports restrictions, buy-back provisions, grant-back clauses,[72] tie-in sales,[73] and tie-out clauses.[74] Some of these restrictions stemmed from stipulations providing KWU with extensive control over the organizational structure, financial function (foreign exchange remittance), and staffing of the joint ventures established with Nuclebrás. The shareholders' agreement over NUCLEN established the creation of a Technical Committee as part of the Technical Directory. This committee—composed exclusively of Germans and one Brazilian observer—had final authority over all design and technical decisions, including the quality of components and services.[75] Explicit quality assurance clauses granted KWU exceptional prerogatives in securing the supply of components and equipment by KWU and associated firms as well as tie-in sales. In essence, Brazil acceded to KWU's near monopoly over imported services and equipment for the first four (out of eight) plants.[76] For the last four plants, however, German supplies would be subject to more competitive standards. Similarly, the requirement of unanimous (or super-majority) decisions provided KWU with a virtual veto power and with the prerogative—through its control of NUCLEN's administrative council and Technical Committee—of overriding decisions by the directorship or attempts by the Brazilians to modify the reactor's design. In turn, the penalties assigned to KWU for errors in conception were minimal, and KWU insisted on limiting its own risk to the initial amount of capital investment.

The pattern of KWU's control of joint ventures with Nuclebrás through the unanimity clause also operated at NUCLEP, the heavy components factory, whose directorship could not veto a decision by the KWU-controlled Technical Committee. This control helped impose limitations on the capability to license, establish technical cooperation agreements, and perform any modifications in the production program.[77] NUCLEP could make free use of the patents and information transferred to it only after the production of the sixth set of components, that is, with the completion of the eighth plant under the agreement, scheduled for 1995.[78]

Restrictions in Articles 3 and 4 of the 1975 agreements made exports contingent on the consent of the supplier.[79] In most cases, the components and processes subject to these restrictions were of a conventional nature and were not subordinated to security-related concerns regarding exports (such as those underlying the London Suppliers Group 1977 guidelines) but to commercial considerations alone. KWU's approval was also required in the signing of any cooperative agreements, according to section 12.5 of the commercial agreements. Moreover, KWU was recognized as sole supplier of technology and equipment (a tie-out clause) for new projects involving NUCLEN.

In the area of uranium exploitation and commercialization, the German firm Urangesellschaft (UG) was ensured the equivalent of 20 percent of uranium reserves, with potential increase to 49 percent, conditional on Brazil's approval. The 20 percent buy-back provision was contingent on Brazil's right to fulfill its own demand first, although that level of domestic demand was quite unlikely. Brazilian critics considered the creation of the joint venture NUCLAM a clear departure from the old policy of national monopoly over mineral exploitation. NUCLAM was allowed to use UG as its exporting agent for the remaining 51 percent but was not required to do so.[80] As in the case of other joint ventures, the Technical Committee had to approve decisions unanimously, granting a veto power to the German partners. NUCLAM was thus an effective instrument to secure West German access to uranium, although Brazil expected this association to pay off in facilitating a marketing network.

The most criticized contractual aspect of the nuclear agreements was the uranium-enrichment process, a responsibility of the joint ventures NUCLEI and NUSTEP. The two were created after Nuclebrás signed a patenting and licensing agreement with the German firms STEAG and INTERATOM, which, in turn, were granted an option to participate in the commercial enrichment plant. The joint ventures in this area were perceived, at the time, as part of a concerted effort by West Germans to develop indigenous fuel-cycle capabilities.[81] Nuclebrás's subsidiary NUSTEP, based in Germany, was committed to share with the licenser, at no cost, any patents or know-how that it might develop, contingent on the approval of all NUSTEP's shareholders; STEAG made a reciprocal commitment vis-à-vis NUSTEP. There were limitations on both sides regarding sublicensing. German risk capital would be maintained at original levels rather than adjusted to eventual industrial-scale production. Brazil's payments would be spread over the first three years following the agreement (with a third down payment). These conditions were not very favorable to Nuclebrás.[82] First, Brazil's very commitment to finance research on the jet-nozzle procedure, an economically unproven technique, involved significant risks. Moreover, critics considered Brazil's agreement to underwrite research performed in West Germany a politically intolerable and economically unjustifiable step.[83] In addition, this uranium-enrichment technique consumed considerably more energy than the alternative procedure of gaseous diffusion, most commonly used by the United States and other nuclear industries.[84] Finally, the effective transfer of technology in this area was contingent on Brazil's completion of the first four power plants.

This last contingency epitomizes the net effect of the minority partner KWU's cumulative control over Nuclebrás's subsidiaries.[85] Buttressed by suppliers' credits, KWU insisted on a program of no less than eight reactors, locking Brazil into a single technical model of pressured water reactors. In contrast, Argentina's plant-by-plant bargaining process allowed sequential decision making (and more diversified learning), potentially reducing long-term risks. Although the establishment of joint ventures may suggest some adaptation by German suppliers to Brazil's preferences, such ventures also offered some protection to Germany against the risk of cancellation of any scheduled plants.[86] Brazil's investment in NUCLEP, for instance, reinforced the commitment to build at least eight plants.

What general lessons can we learn from this experience, regarding the utility of joint ventures in high technology for recipients? Foreign investors often prefer joint ventures when their contribution of venture capital is reduced (as in uranium enrichment), when they seek access to local markets for factor inputs (in this case, uranium) or sales of final products (reactor equipment), and when there are political benefits (Brazil's special relationship with Germany).[87] In such cases, they may use a minority joint venture in combination with direct purchases, licensing, and technical assistance to obtain effective control, at times resembling a wholly owned foreign subsidiary.[88] These conditions can limit learning by the recipient and may overlook national technical needs insofar as effective technology transfer considerations become extraneous to the operation of the joint ventures.[89]

These dangers notwithstanding, joint ventures based on foreign capital, technology, and management raise the vendor's stake in the venture's success and the product's quality.[90] They may also facilitate a rapid transfer of relatively advanced technology.[91] The transfer of effective control to the recipient or host state is usually contingent on the level of local technical competence. In general, a consistent parallel investment in domestic capabilities and an appropriate implementation of technology imports restrictions tend to increase the benefits associated with any technology transfer mode, not just joint ventures. It is even possible "to get more experience from a turnkey project with a big training component than from purchasing the various technological elements separately."[92] In other words, even turnkey projects can be transformed into "grey boxes," as the Argentine nuclear program suggests. These contrasts support the assertion that blanket policies for or against specific modes of technology transfer may not always be useful and that almost any transfer can be exploited as a building block to proceed up the technological ladder. The contractual mechanism may determine the structure of technology transfer but leave enough room to maneuver in the implementation.

Overpricing of technology imports is also branded as evidence that recipients are disadvantaged in bargaining for technology.[93] The following overview illuminates some of the difficulties involved in assessing prices and costs. The original estimates of Brazil's 1975 nuclear program projected a total expenditure of about $18 billion (in 1975 U.S. dollars), distributed as follows:[94] $13.8 billion for eight power plants, $3.7 billion for the fuel-cycle facilities,[95] and $900 million for investment in technology.

Had the agreements remained in place (a counterfactual), by 1983 the total cost would have risen to between $24 billion and $35 billion (including indirect costs), or double the original estimates.[96] In reality, over $8.3 billion was spent by 1990, and an additional $7 billion was required through the end of the decade ($5 billion for the completion of Angra 2 and 3 only).[97]

The original estimated cost of Brazil's first plant under the agreement (Angra 2, expected to operate by the late 1990s) was $1.5 billion (in 1975 dollars), considered well beyond the international market price at the time (about $600 million, or $500 million in Germany itself).[98] The 1979 congressional investigating committee on the nuclear program found some justification for the added technology costs, commissions for purchases in Germany, and other adjustments, but found the costs per plant overpriced nonetheless, by about 20 percent (or $144 million).[99] Interestingly, Newfarmer's study of the International Electrical Association as a cartel found that the consequence for recipient countries (particularly less developed ones) from the association's operating procedures was an over-pricing of about 20 percent.[100]

The cost of Argentina's first plant (Atucha 1) was $241 million in 1974; the second plant (Embalse) was $1.3 billion, and the third plant (Atucha 2), still under construction in 1994, has climbed above $4 billion.[101] The price gap between Argentina's and Brazil's first plants clearly dissipates with the others. In Argentina's case, however, CNEA's national content requirements undoubtedly added significantly to the total costs of each plant. Moreover, considering Argentina's overall energy balance, its nuclear plants (and Brazil's as well, some might argue) were far from cost-effective, particularly relative to hydroelectric plants.[102] Yet, although an important topic in itself, the economic analysis of the nuclear programs in light of overall energy alternatives is beyond the focus of this book.

Another useful measure of comparison among nuclear plants is their cost per kilowatt (KW). In these terms, Angra 2's estimated cost was over $2,500 per KW. Argentina's Atucha 1, completed in 1974, was $1,300 per KW, but Atucha 2 was expected to reach over $3,000 per KW, owing to delays and consequent financial costs.[103] Spanish, French, and German plants were built at less than $1,000 per KW. Naturally, technical and size differences among these plants matter when evaluating costs.

From a more technological angle, one might evaluate costs by studying their distribution among engineering services, capital goods supplies, assembly operations, and personnel training. Such disaggregation could arguably measure the extent to which costs included the transfer of effective technological know-how and not just the purchase of equipment.[104] Unfortunately, no authoritative source can guide this effort, and only some overall trends are discernible.

First, equipment costs in Brazil accounted for $1.8 billion of the total expenditure of $2.9 billion by the early 1980s.[105] Engineering costs for the first two plants (including both KWU and NUCLEN) were estimated at about $400 million by 1981.[106] Technical training for engineers up to 1985 absorbed a total of $275 million.[107] Capital goods imports were clearly the main form of incorporation of foreign technology, reaching 73 percent of all electromechanical supplies by 1981.

Second, the proportion of imported equipment was to be much higher for the first four plants, and the nationalization index would rise only after the fifth plant. Therefore, the program's first years would involve heavy foreign exchange expenses (higher than 50 percent of the costs), whereas the less costly stage (fifth plant onwards) would not take place until the first four plants were completed. For comparative purposes, the proportion of outlays in local currency for Argentina's second plant (Embalse) was 68 percent.[108]

Third, in rough terms, Brazil's annual payments for imported R&D (not including capital goods) were between $25 million and $60 million. Domestic R&D efforts absorbed about $17 million, combining both the Nuclear Energy Commission and Nuclebrás's Center for Development of Nuclear Technology (CDTN). At worst, therefore, these figures suggest that the ratio of foreign to domestic R&D effort was about 3:1. This ratio characterized other industrial sectors as well and reflected no particular dedicated effort by Brazil in the area of R&D. At best, foreign to domestic R&D expenditures represented a ratio of 1.5:1.[109] An inference from some of Argentina's reported figures on the nuclear program (for 1985) reflects a ratio approximating 1:10 in foreign to domestic R&D costs.[110] This ratio is comparable to that of developed countries—which normally invest about ten times more on domestic than imported R&D across sectors—and reflects CNEA's effort to integrate Argentina's scientific and technological community into the nuclear program.

Finally, the training of NUCLEN's personnel by KWU in Erlangen (West Germany), per year and per trainee, was almost three times the cost of training a Ph.D. at MIT.[111] The total costs of training up to 1985—$275 million for about five hundred trainees—imply a per capita cost of over half a million dollars. This might compare negatively with the domestic costs of training through, for instance, the National Research Council (CNPq) or CNEN, although different types of training are at work, as I analyze in Chapter 7.

No discussion of the cost of nuclear programs in these countries can end without a reference to their place in the overall external debt of two of the industrializing world's largest debtors. Brazil's nuclear program was estimated to account for $7.4 billion, or over 5 percent of the national debt, in 1989.[112] Argentina's nuclear foreign debt represented $2.3 billion in 1988, or 4 percent of its $55 billion foreign debt, although some estimates place that debt at over 7 percent.[113] Both programs thus contributed significantly to foreign indebtedness, while siphoning resources that might have been available for alternative sources of energy. I return to the technological angle of gains and costs (including opportunity costs) in Part III.

In 1979 Brazil's congressional committee investigating the nuclear program called for changes in the original agreements, including the reduction of the total number of plants, the elimination of clauses making technology transfer contingent on the completion of all plants, the adjustment of German risk-sharing responsibilities, and the restructuring of the composition of technical committees in joint ventures.[114] The committee's chairman at the time was Senator Itamar Franco, who later assumed the presidency of Brazil following the impeachment of President Collor in 1992. In 1985, with the country's return to democracy, Brazil's parliament approved Legislative Decree No. 3, demanding parliamentary approval of all provisions, protocols, contracts, or acts in relation to the nuclear agreements between Brazil and Germany.[115] The changes that followed the original bargain of 1975 render greater support to the theory of "obsolescing bargain" than to its counterpart; the latter predicts a growing— rather than a weakening—bargaining advantage to multinational corporations, relative to that of host states.[116]

The International Context: A Limiting Factor?

This comparative analysis of bargaining in nuclear technology yields some general lessons, applicable to a broader industrial spectrum and to a wider set of recipients and suppliers.

First, external constraints—in the form of international political, market, and financial factors—provide an important backdrop against which domestic choices are made but certainly do not determine those choices. During technology search and negotiations, latecomers can maximize the so-called advantages of backwardness.[117] Despite differences in performance and outcomes, the experience of Argentina and Brazil seems to suggest that national objectives are not always compromised as a result of international pressures. The ability of the two countries to extract concessions from suppliers in fuel-cycle-related technologies is a most tangible expression of a recipient's leverage. The particular nature of the international nuclear market may limit, but only to some extent, the generalizability of findings from this particular study. Findings will be mostly applicable to sectors characterized by long lead times and relatively small markets of a zero-sum nature, where each transaction involves high financial stakes and where fierce competition among suppliers adds up to an advantageous bargaining position for recipients.[118] The overall difficulty that latecomers may face in their attempts to break into highly dynamic, high-technology industries remains.[119]

Second, bargaining advantages cannot be deduced simply from international structural conditions. Identifying the respective win-sets requires prior knowledge of levels and content of macropolitical consensus and bureaucratic autonomy, both of which shape the bargaining context in a very fundamental way.[120] Thus, a state with lower levels of consensus over industrial policy may be able to extract greater concessions from foreign suppliers than one with a more coherent industrial strategy. A sectoral agency endowed with bureaucratic autonomy in the midst of low macropolitical consensus can narrow the size of the win-set to its own institutional preferences, thus lowering uncertainty about the investment.[121] That confers some bargaining advantages and can increase the negotiators' leverage over the distribution of benefits from the international bargain. These conditions may neutralize otherwise important considerations such as the size of the domestic market or a state's overall power capabilities. Knowledge of macropolitical consensus and bureaucratic autonomy helps investors estimate the transaction costs that might be involved in the bargaining process.

Third, the domestic context explains the particular zeal of Argentine negotiators in maximizing national entrepreneurial and technological resources. Integrating technically sophisticated officials representing a sectoral perspective into the bargaining process improved CNEA's ability to counter the oligopolistic practices of technology suppliers and to resist inappropriate technology transfer mechanisms and onerous conditions.[122] The dominant political concerns of Brazil's negotiators—and not the absence of qualified scientists and technologists in Brazil—explain many of the characteristics of Brazil's experience with technology transfer in the nuclear sector. In other words, neither idiosyncratic individuals nor functional groups can explain differences in behavior and outcome. Rather, the structural and institutional conditions within which the programs originated shaped that variation. Knowledge about the content of macropolitical consensus and about the institutional imperatives of a state agency (i.e., knowledge about the distribution of domestic "structural power") can improve our ability to predict actual outcomes from potential "bargaining power."[123]

Fourth, the strategy underlying CNEA's bargaining included the attempt to disaggregate sources of supply. Thus, Argentina did not sign a single comprehensive agreement in the nuclear sector with an external supplier but a series of differentiated packages for separate plants and fuel-cycle facilities, all carefully selected in bids guided by a consistent set of criteria. Brazilian negotiators' emphasis on the accelerated completion of nuclear power plants for electricity generation explains—but only to some extent—a willingness to accommodate a single supplier of plants and fuel-cycle technology. International political pressures exerted themselves to narrow the pool of suppliers, eventually to KWU and its partners. These conditions weakened an otherwise strong bargaining position on the part of Brazil and helped the technology supplier implement its preference for no less than eight power plants, for an unproven uranium enrichment technology, and for significant control of the joint ventures created as a result of the agreements. This point suggests that diversification of technology sources is clearly an important instrument for recipients to counter special clauses and overpricing.[124] Moreover, it reinforces the notion that a privileged bargaining position (potential) is not always translated into an optimal—or even an effective— (actual) bargaining outcome.[125]

Fifth, multinational firms providing both equity and technology to joint ventures with national enterprises find ways to maintain managerial control of such ventures, even where they are minority partners.[126] With managerial control, they are better able to maximize imports of their own equipment at the expense of domestic equivalents and to impose strict contractual provisions (special, or restrictive, clauses). Suppliers' credits are an additional avenue to secure their own imports, but their effect can be countered. A recipient's effort to obtain independent financing is no guarantee against overpricing, as the case of Argentina's Atucha 2 suggests. Overall, the "obsolescing bargain"—and host states' ability to undo some of the initial unfavorable arrangements—appears to hold for nuclear industries as well.

Sixth, both Brazil and Argentina opted for foreign reactor technology, as has been the case for most nuclear programs in the industrial world.[127] The characterization of Brazil's program as "technologically dependent" and Argentina's as "self-reliant" is therefore inappropriate. There are vast discrepancies in the literature as to what defines a nuclear program as more or less "dependent" or "self-reliant." In some cases absolute mastery of the nuclear fuel cycle is the benchmark, yet both countries aimed equally at acquiring that capability. For others, reliance on a diversified pool of technology suppliers—as opposed to a single one—is taken as a sign of independence. Neither case makes the recipient *technologically* independent, although multiple sources of supply can reduce, as we have seen, a recipient's commercial and *political* vulnerability.[128]

The importance attached to technological considerations in Argentina was evident in CNEA's efforts to integrate its national scientific and technological human power into the program. As in other areas, the contrast with Brazil was quite striking, and it is to this dimension that I turn now.

CHAPTER 5

The Scientific Community: "Insiders," "Outsiders," and Counterfactuals

Já é difícil ser profeta do futuro, ainda mais . . . [saber] onde estaríamos se tivéssemos prosseguido [o projeto do tório].[1]

(It is hard enough to be a prophet of the future; it is even harder . . . [to know] where we would be, had we pursued [the thorium project].)

This chapter analyzes the role of the Brazilian and Argentine scientific communities, particularly physicists, in their country's nuclear programs.[2] That role was largely influenced—in each case—by the structural and institutional conditions identified so far. Contrasting degrees of macropolitical consensus and bureaucratic autonomy in Brazil and Argentina resulted in two distinctive patterns of state-scientists interactions. The next section explains why Brazil's scientific community was essentially marginalized from the large-scale program begun in 1975.[3] The chapter then analyzes this community's advocacy of an alternative industrial policy and nuclear program, one that was rejected and became a historical counterfactual. A third section explains why segments of Argentina's scientific community were attracted to the nuclear effort and became among its most fervent advocates. The chapter next examines the implications from both cases for the study of scientists in an industrializing context and ends with some general conclusions from these cases for the generic study of state-scientists relations.

High Consensus, Low Bureaucratic Autonomy, and Brazil's "Outsiders"

Brazil's scientific community was largely excluded from the initial design and development of the country's nuclear industry. This exclusion is quite striking in light of three major considerations.

1. The technical caliber of this community was high. Brazil's scientific capacity in nuclear physics was globally recognized and, some argue, had historically surpassed that of Argentina.[4] Prominent visiting physicists—David Bohm, Richard Feynman, J. Robert Oppenheimer, C. N. Yang, Eugene P. Wigner, Emilio Segrè, and Guido Beck, among others—trained Brazilian physicists in those early years. Several among the latter group achieved world recognition, notably César Lattes, José Leite Lopes, Mário Schemberg, Marcelo Damy de Souza Santos, José Goldemberg, Oscar Sala, P. A. Pompeia, and Elisa F. Pessoa.[5] César Lattes, internationally known for his work in cosmic radiation, founded the Brazilian Center for Physics Research (Centro Brasileiro de Pesquisas Físicas) in 1949, which became a large research institute with the greatest number of theoretical physicists in Latin America, mostly trained in first-rate U.S. academic institutions.[6]

2. In industrialized and developing countries alike, important functions have been traditionally bestowed on physicists (including theoretical physicists), particularly in the context of applied technology policies.[7] In the second half of the twentieth century, physics epitomized the paradigm of science in modern societies, embodying expertise and competence.[8]

3. Brazil's post-1964 regime was characterized as one permeated with a scientificist-technocratic ideology.[9]

The exclusion of Brazilian scientists, of course, seems far less puzzling when we consider the political and institutional sources of Brazil's nuclear policy analyzed in Chapter 2. As argued, the program designed in 1974–75 was compatible with the core macropolitical objectives of the coalition that held power between 1969 and the mid-1970s. The possibility of slowing down the development of a nuclear industry—by embracing a different program advocated by Brazil's organized scientific community—was regarded as just such a departure. The more gradual nature of the alternative advanced by prominent scientists involved perceived risks of energy undersupply, delays, and capital and technology shortages; such an alternative program was thus outside the domestic win-set guiding Brazil's bargaining with West Germany and KWU. The maintenance of Brazil's political-economic model had no place for attempts to "reinvent the wheel," in the words of senior governmental officials.[10]

In other words, efficiency and immediacy of results were the criteria that guided technocrats and managers. Scientists, in turn, were concerned with opportunities for involvement in the development of new technologies. When the United States entered the nuclear field, it was in a far better position to reconcile those two considerations. In contrast, the availability of extant nuclear technology worldwide allowed technocrats among latecomer countries (although it did not necessarily force them) to forgo a meaningful input from their own national scientific community. Such technocrats argued that France, West Germany, and Japan had imported nuclear technology as Brazil did and that the development of a nuclear industry was much less associated with scientific skills than with engineering and economic criteria.

The organized scientific community espoused a strategy they characterized as "greater technological self-reliance," which would increase the community's own participation in the program. I analyze this strategy in greater detail in the next section. The community's emphasis on integrating Brazil's industrial entrepreneurs into the program was not only self-interested (it had the promise of maximizing the use of local scientific-technical resources) but also compatible with the overall industrial policy advanced by opponents of the regime's political-economic model of industrialization, heavily represented in the academic community as a whole. The nuclear program thus became a centerpiece in the broad political debate over socioeconomic development and stood at the core of a growing movement aimed at dismantling that model. Scientists played an important role in coalescing their rank and file, prominent industrial entrepreneurs, and the media in their opposition to military rule.

The dominance of central political-economic guidelines (in tune with the model's requirements) over sectoral interests curtailed radical departures from the technical and industrial blueprint advanced by decision makers. Nuclebrás reflected this technocratic consensus, and a vigilant crew of military officers with little substantive expertise and a great deal of mistrust for academic experts supervised its implementation.[11] The regime's prevailing approach permeated CNEN as well, particularly during the late 1960s and early 1970s; CNEN played a marginal role in designing the nuclear sector and in negotiating the agreements with KWU.[12] CNEN's bureaucratic autonomy had been phased out in 1967 with its transfer from direct subordination to the presidency to effective accountability to the Ministry of Mines and Energy. Earlier on, during the early 1960s, CNEN had enjoyed a high degree of lateral autonomy as a federal *autarquia* with administrative and financial autonomy, and with prerogatives to negotiate treaties, financing, and sale of feasible materials. This was a period of little macropolitical consensus (under President Quadros), when CNEN was able to advocate firmly what were, at the time, its own institutional preferences—heavy water and thorium cycles—with strong support from scientists and industrial entrepreneurs.[13] By the late 1960s, macropolitical consensus and a segmented decision making led to a program harmonious with a political-economic strategy emphasizing state entrepreneurship and foreign technology.

The nature and degree of macropolitical consensus and the limited sectoral bureaucratic autonomy thus precluded a widespread incorporation of Brazil's scientific community into the nuclear program.[14] Nuclebrás had about 250 high-level technical professionals and less than 200 mid-level technical employees at the inception of the program in 1975, ready to begin construction of Brazil's second nuclear power plant.[15] In contrast, Argentina's CNEA relied on at least 2,500 scientists, engineers, and technicians while negotiating the supply of its first plant and on over 4,000 professionals for the second plant.[16] Not only were Brazil's academic scientists excluded; the program also left out governmental nuclear research centers, such as the Institute of Atomic Energy of the University of São Paulo (housing about half of the nuclear science community in Brazil), the Institute of Nuclear Energy in Rio de Janeiro, and the Institute of Radioactive Research in Belo Horizonte.[17] A small group of "insider" scientists with official government ties did join the nuclear program.[18] Yet most of the academically prominent and politically visible members of Brazil's scientific community remained "outsiders," as did a large number of academic scientists.[19] Their exclusion strengthened the ability of leading members of scientific societies to build institutional support for opposing the government's nuclear program.[20] Brazil's Physics Society and the Brazilian Society for the Advancement of Science, as well as members of other professional societies (physical, natural, and social scientists), were particularly active in articulating the opposition to the 1975 agreements.

Criticisms on the basis of the technical and political-economic nature of the program were symptomatic of a general estrangement between the regime installed in 1964 and the organized scientific community. The beginnings of a science policy in Brazil has been associated with the creation of the National Research Council (Conselho Nacional de Pesquisas—CNPq) in 1951 by Admiral (physicist) Alvaro Alberto. In essence, the inception of a science policy and a nuclear policy were concomitant. Although the Brazilian Center for Physics Research began obtaining support from CNPq, it never fully entered the area of nuclear energy. Many Brazilian scientists blamed this divorce between their theoretical endeavors and the country's technological requirements on the inability of the state to create industrial demand for local scientific output. In fact, in the development of Brazilian physics from its beginning until 1982, there was only a single instance in which a group of physicists was called upon to use their expertise directly in the service of the state.[21]

Since 1964 the more or less autonomous scientific institutions and universities became centers of debate, challenging the fundamental prevailing political and economic order. As one of the best organized sectors within the scientific community, physicists had developed a long tradition of political involvement.[22] Early Brazilian physics was influenced by European scientists who fled totalitarian regimes.[23] The international political context of physics (Einstein, the atomic bomb, the Oppenheimer affair) exacerbated the politicization of this community, as did Brazil's grim poverty and distributional inequities. Under the repressive environment that followed the military coup, many among the country's most prominent physicists—including the president and vice president of Brazil's Physics Society, the scientific director of the Center for Physics Research, José Leite Lopes, and Mário Schemberg, José Goldemberg, and Jayme Tiomno, among others—were either expelled from their research institutions or became subjects or harassment by the internal security and information arms of the regime, particularly following the infamous Institutional Act No. 5 in 1969.[24] The main research groups at Belo Horizonte's Institute of Radioactive Research (University of Minas Gerais) were also disbanded. Nuclear research was considered too sensitive to be left to the opposition, and there was an evident effort to limit the physicists' political role.[25]

In contrast to Merton's (1973) generic characterization of scientific communities, most of these scientists were not alienated from the "subordinate strata in society." There was an unresolved tension within their *Weltanschauung* between adherence to the tenets of the Republic of Science (science for science's sake) and an instrumentalist view of science in the service of socioeconomic and political problem solving.[26] Thus, several critics of the nuclear program became more intensely involved in energy policies in general, and engaged in research on alternative (non-nuclear) sources, mainly biomass, and the alcohol program, based almost entirely on indigenous research and development.[27] A summary statement of the basic consensus guiding their political activism is included in a memorandum following the 1975 General Assembly of Brazil's Physics Society. The document attacked economic policies detrimental to the development of a national technological capability and demanded participation in the political debate over technological options, as well as integration of nuclear research institutions with the university system, a policy of strict state monopoly over natural resources of strategic importance, investment in a long-term, autonomous program of research and development in future reactors (high temperature, breeder, and thorium-based reactors), and opposition to military uses of nuclear technology.[28]

The most important institutional opposition to the regime, around which physicists and other scientists organized their political activism, was the Brazilian Society for the Advancement of Science (SBPC). Established in 1948, at the time of a prevailing atmosphere of hostility to basic science among state governmental circles (under Governor Adhemar de Barros), the SBPC was founded partly to address the need for combining basic and applied science in a harmonious way.[29] Social scientists became active participants around 1968. From the early 1970s on, the annual meetings received wide press coverage and became a forum for debate on national issues, severely criticizing the regime's socioeconomic policies. Following the period of heaviest repression, the SBPC emerged as the first institution in civil society to venture into broad political debates. The prominent role of physicists within the SBPC in these years was evident from the programmatic decisions within Brazil's Physics Society regarding the SBPC's activities.[30]

The SBPC's 28th annual meeting (1977) opened its doors to students, thus inviting governmental retaliation and the attempt to cancel its state funding. Instead, this response revitalized and radicalized the activities of the SBPC.[31] Virtually the entire scientific community, including its more conservative elements, protested what it regarded as governmental ambivalence toward scientific and technological development. With the participation of more than ten thousand people, the meeting became a political event cementing an informal alliance among scientists, lawyers, and members of the clergy and the press, and advancing the need to reach out to broader societal segments and to help articulate the right of other groups to political expression. Panels on nuclear energy, the physicists' centerpiece in a comprehensive criticism of the government, attracted over three thousand people. The SBPC had become "one of the few tribunes of free debate in Brazilian society."[32]

In sum, those who shaped Brazil's nuclear program underestimated the ability of the scientific community to turn its own exclusion into an effective instrument of political mobilization against the regime. This community emerged remarkably resilient from the attempt to eliminate it as a political force and succeeded in exploiting a deteriorating macropolitical consensus within the ruling coalition brought about by the collapse of the so-called economic miracle in the mid-1970s. By the early 1980s, neither the social costs of the economic boom nor the severity of the external debt could be disguised. Backing the demands of important sectors of the national bourgeoisie, the scientific community joined in the frontal attack against *estatização* and reliance on foreign capital and technology. Yielding to this cumulative pressure, the regime attempted a reconciliation in the early 1980s, by reversing some of its policies regarding nuclear research.[33] By 1983 CNEN was actively developing a Brazilian-designed research reactor, in cooperation with universities, research institutes, and private firms.[34]

At this juncture, structural and institutional conditions had come full circle. Low macropolitical consensus and a renewed bureaucratic autonomy for the nuclear sector had created the setting for policy change. The government's "Plano 2000" announced a partial freeze on the implementation of the 1975 agreements. The whole fabric of state-scientists relations underwent a radical transformation, with political liberalization and the return to democracy in 1985 bringing prominent scientists into important state and federal positions.[35]

The Program That Never Was: Analysis of a Counterfactual

The main criticisms of the 1975 agreements advanced by leading scientists centered on the regime's preference for a technical option (light water/enriched uranium) that was presumed to imply greater technological dependence and a limited role for national experts. Instead, scientists advocated an alternative path (heavy water or thorium cycles), which they expected would employ more local scientific and technological staff and use more natural local resources (uranium, thorium). Opponents of the 1975 agreements were able to buttress their critique with additional claims of bargaining failures—outlined in Chapter 4—most particularly Brazil's acceptance of the jet-nozzle enrichment technology, one not yet proven on an industrial scale and with little prospect for an effective technology transfer. This arrangement, scientists argued, opened the possibility of exchanging one type of energy dependence (enriched uranium) for another (oil).[36]

We can analyze the "national" nuclear program advanced by Brazilian scientists as a counterfactual: the kind of program that would have resulted from alternative combinations of macropolitical and institutional conditions. As argued, such conditions indeed led to a program compatible with the preferences of the organized scientific community in the early 1960s, a program dismantled after 1964. Scientists overwhelmingly favored a nationally developed natural uranium/heavy water or thorium technology based on incipient research but capable in the long run, they thought, of ensuring authentic national scientific development.[37] This path reflected a concern with the diffusion of knowledge from basic to applied research and with the integration of research efforts with industrial technological development.

Since 1965 a group of over 50 physicists and engineers at the Institute of Radioactive Research (Belo Horizonte), known as the "thorium group," had been involved in designing a natural uranium reactor that could later operate with thorium, a fissile material abundant in Brazil.[38] The departments of materials and metallurgy of the Institute of Radioactive Research cooperated in this project as did the Rio de Janeiro Institute of Nuclear Engineering. Components for the reactor were to be produced almost entirely in Brazil. A Brazilian firm was asked to study pressure vessel design. An agreement with France established cooperation in research and design of this reactor concept. The group chose heavy water technology because it provided the reactor with flexibility to operate with three fuel mixtures, thus securing Brazil's independence of fuel supply. Since heavy water does not involve secrets of production (unlike uranium enrichment, necessary for light water reactors), they argued that it ensured independence from the nuclear powers.[39] The notable fact about the thorium group was that it was a worldwide pioneer in researching this technology.[40] It managed to build, in cooperation with local industry, a subcritical assembly (not a reactor) called CAPITU (Assembly Heavy Water, Thorium, Uranium).[41] In 1969 CNEN discontinued its support for this group, together with others working on natural uranium cycles in other institutions. This followed the formal decision to opt for light water/enriched uranium plants, and CNEN's alignment with the regime's consensus.[42]

Following the 1975 agreements, Brazil's Physics Society began promoting a proposal for a ten-year program that, by some estimates, would have required an investment of $10 to $20 billion.[43] Director of São Paulo University's Institute of Physics and president of the Physics Society, José Goldemberg argued that many scientists and industrialists would have preferred "a more modest program in which they could play a dominant role while acquiring the necessary technology."[44] Technocrats in charge of the nuclear program contended that there was no critical mass of scientists and engineers in Brazil to sustain a more self-reliant nuclear program that would guarantee fast delivery.[45] On the one hand, the Physics Society reported over 277 Ph.D.'s in physics and 417 M.Sc.'s in Brazil in 1975.[46] About 50 researchers in all areas of physics were added annually to the academic ranks, although the 1976 Second Basic Plan for Scientific and Technological Development emphasized the shortage of physicists for higher education needs, projected at 700 Ph.D.'s and 1,500 M.Sc.'s by 1979.[47] On the other hand, the backbone of an indigenous program would have required greater numbers of engineers than of physicists, or a ratio of about 15:1.

In response, a report by the Physics Society argued that the model adopted in the 1975 agreements shaped the nature of the personnel it would require. Thus, a program resembling those of highly developed countries, such as Japan, West Germany, the United Kingdom, and France, would demand 2,000–2,500 scientists (two-thirds of which would be Ph.D.'s and M.Sc.'s), including about 500–600 physicists. The alternative program model—resembling those of industrializing countries like Spain and Turkey—would require 1,400 scientists, including about 200–300 physicists.[48] The report recommended the first model for a (long-term) program aiming at meaningful absorption of technology and better coordination between academic programs at major universities and Nuclebrás's activities. Scientists blamed the lack of progress in enrichment technology, for instance, on the failure to attract serious academic physicists[49] and pointed to the successful absorption of Brazilian scientists in the aerospace and arms industries.

Evaluating the potential merits of Brazil's counterfactual nuclear program requires the control of too many technical and political variables to make it a comprehensive exercise. However, to some extent, the path not taken can be assessed by analyzing the role played by the scientific community in the Argentine program, which resembled many of the characteristics favored by Brazilian scientists.

A Maverick Amidst Macropolitical Chaos: Argentine Scientists as "Insiders"

As argued in preceding chapters, the peculiar structural and institutional features surrounding nuclear decision making in Argentina led to the adoption of a nuclear program aimed at including not only private industrial entrepreneurs but large numbers of scientists and technologists as well. CNEA's (and the Navy's) outstanding degree of bureaucratic autonomy from civilian and military regimes, and its resulting institutional stability, allowed it to promote clientelistic networks from among this community. The inclusion of experts in effective planning, negotiations, and implementation strengthened CNEA's autonomy even further, enabling it to pursue, in a sustained fashion, its built-in political and technological imperatives.

Bureaucratic autonomy kept CNEA's policies insulated not only internally but also externally, aloof from broad societal debate. Yet the agency used public tactics designed to highlight the convergence of state and societal preferences by appealing to shared symbols such as "self-reliance" and deference to official expertise.[50] The program's success was often exaggerated, as an instrument in the intraservice struggle among the armed forces and in the attempt to induce and reward clients. In fact, Argentina's nuclear program, like Brazil's, relied on foreign technology. There was a genuine effort, however, to absorb technology effectively by recruiting scientists in related research and development activities. The selection of a heavy water path aimed at facilitating the integration of national resources, under the assumption (perhaps erroneous, in retrospect) that heavy water production required lower levels of technological complexity compared to light water/enriched uranium cycles.

The beginnings of Argentine physics are paradoxically associated with a fraudulent attempt in 1951 by an obscure Austrian physicist with delusions of grandeur—Hans Richter—to convince President-General Perón that he could produce a hydrogen bomb. Despite a very expensive laboratory set up for Richter's work in Bariloche, evidence that the program was a fiasco—provided mostly by prominent members of the Argentine Physics Society in 1952—resulted in Richter's dismissal. Following this incident, Perón established the Argentine Atomic Energy Commission (CNEA), which began hiring the first Argentine physics research group. This time, the emphasis was on scientific competence, not political loyalty, to ensure the acquisition of an effective capability to master the nuclear cycle, from nuclear fuel and nuclear reactors to reprocessing and enrichment. CNEA established the Balseiro Institute of Physics in Bariloche, with full-time teaching personnel, scholarships, a modern research environment, and strong ties to European and U.S. institutions. The Institute produced several Ph.D.'s in theoretical physics, developed some branches of experimental physics with high applied content (such as chemical physics, atomic physics, and physical metallurgy), and gained an international reputation as the best school of physics in Argentina.[51] In Brazil the tradition of scientific academic training *preceded* the creation of the Nuclear Energy Commission (CNEN) and remained independent from it.

CNEA protected the Balseiro Institute, which became an island of stability throughout a long succession of authoritarian and democratic political regimes.[52] Its research activities, even in theoretical subjects, were seen as compatible with the "applied" outlook of successive regimes interested in the development of a national nuclear industry. The stability and protected environment of CNEA was a significant source of attraction for Argentine scientists, who became effective partners and ardent supporters of CNEA's objectives and strategies.[53] In contrast, scientists at the universities were more often objects of repression than those engaged in nuclear energy programs and their associated research institutions. The emphasis on professionalism by CNEA's first president, Admiral Iraolagoitía, contrasted with Peronist purges of scientists (many of whom were his political opponents) at the universities. The strongest blow to Argentine physics research (unrelated to CNEA's work) came under Onganía's military rule (1966–68), particularly after the repressive incident known as "the night of the long sticks." At least one hundred physicists, including entire research teams, left Argentina for Chile, Venezuela, Peru, the United States, and Western Europe. Over half of the

science faculty of the University of Buenos Aires (UBA) and 90 percent of the physics department submitted their resignations.[54] Scientists did not fare better during the Peronist interlude of 1973–76.
This time over 50 percent of the (remaining) leadership of Argentina's
Physics Society fled the country. The association ceased to exist until 1981; exiled Argentine physicists accepted respectable positions
in European and North American institutions. Repression and tor-
ture following the military takeover of 1976 were accompanied by
the dismissal of nearly one hundred research scientists supported by
the National Council of Scientific and Technological Research, over
six hundred from other governmental research institutions, including the National Physics and Technology Institute, and a few at
CNEA itself.[55]

This review of the development of Argentine physics highlights
the marked contrast between the disruption of university activities
(mostly in theoretical physics) by frequent political turmoil on the
one hand, and the consistent governmental support for efforts within
CNEA and at the Balseiro Institute on the other. Despite the exile of
a few prominent Brazilian physicists and the harassment of many
others who stayed, the physical elimination and brain drain that took
place in Argentina had no parallel in Brazil. In fact, the regime in
Brazil tolerated the activities of the vocal group I discussed in the
preceding section as one of the few remaining voices of opposition to
military rule.[56] There was a more pronounced duality in Argentine
state-scientists relations, ranging from co-optation to near physical
annihilation. Toward the end of the repressive authoritarian era,
CNEA itself attempted to restructure the physics community in Argentina, an endeavor made increasingly difficult by the agency's progressive loss of autonomy and budgetary resources.[57]

Science, Technology, and Development

My review of the political battles fought by Brazilian scientists in
the context of the country's nuclear program highlights their attempt
to extricate themselves from marginality. Large segments of the scientific community traced their predicament to conditions of international structural dependency, arguing that import-substitution-
industrialization had reinforced a skewed income distribution toward urban middle and upper classes. According to this view, these
elites dominated consumer goods markets and demanded products

similar to those available in industrialized countries. Local producers, in turn, relied on technologies supplied by multinationals—directly or through licensing agreements—at low relative cost and high reliability. This failure to create demand for locally developed technologies weakened scientific institutions, thus alienating them further from productive activities.[58] Science in developing countries, they argued, became a consumption item, whereas in industrialized countries it was an investment item, or part of an organic relationship between technical-scientific and productive growth. The experience of the thorium group was often held as a primary example of the Sisyphus-like nature of scientific development under conditions of structural dependency.

In this scheme of things, the bureaucratic-authoritarian regimes installed in the 1960s were depicted as the quintessential handmaidens of scientific underdevelopment. As evidence of this scientific regression, some pointed to the overall record of the post-1964 era in Brazil and to the brutal repression of the post-1976 Videla regime, which decimated the ranks of the academic scientific community in Argentina.[59] In reality, these regimes' predecessors could not claim a more impressive performance. Brazil's Center for Physics Research (CBPF) was created in 1949 as a way of protecting basic science from the attacks of populist governor Adhemar de Barros. The government of President Café Filho curtailed the work of Admiral Alvaro Alberto, who had been instrumental in the creation of the Center and of the National Research Council (CNPq).[60] Populist President João Goulart (1961) practically dismembered the Center, failing to maintain federal support for scientific institutions.[61] In fact, the Quadros-Goulart (1960–64) era witnessed a dramatic deterioration in funding and an accelerated brain drain. This pattern closely resembles the Argentine experience throughout a succession of unstable populist and military-bureaucratic regimes.

In contrast to structuralist perspectives on the barriers to scientific development, others have blamed the lack of demand for science in developing countries on the inability of academic institutions to produce "applied-minded" scientists. This view is largely associated with the technocratic approach of bureaucratic-authoritarian (particularly military) regimes, which readily enlisted engineers and economists in problem-solving capacities. Emphasis on technical decision making was assumed to give an appearance of formal, rational, "scientific" quality to governmental policy. To ensure unchallenged governmental control of research, the university structure was often bypassed through the creation of new research centers firmly controlled by the state and dissociated from political activity.[62] As Schwartzman (1978:61) notes, different conceptions of science, technology, and development underpinned the confrontation between the techno-scientific and planning bureaucracy linked to an action-oriented engineering tradition and the scientific community anxious to rescue itself from its own marginality. The regime's anti-intellectual tone and depreciation of the theorist were transparent. However, as we have seen, the applied bias transcended otherwise different political-economic regimes. The contention that governments in developing countries are inimical to science (particularly basic science) was by no means unique to the dependency perspective. Traditional analyses of modernization and the role of intellectuals in developing areas shared this assumption, albeit using cultural obstacles to explain it.[63]

There have been attempts to link the normative structure and social context of research to scientists' socio-political attitudes. Theoretical scientists, for instance, have been associated with a more "liberal" outlook, "intellectually disposed to favor innovation and reform."[64] Pure science—more than applied science—has been found more compatible with the scientific ideals of rationality, universalism, individualism, communality, and disinterestedness.[65] Applied scientists as well as engineers, in contrast, are perceived as more conservative.[66] Although this chapter's analysis of Brazil suggests that most of the *politically visible* scientists were indeed theoretical nuclear physicists, all categories were active, particularly institutionally, through Brazil's Physics Society and the SBPC.[67]

Conclusions

The bulk of the literature on science, technology, and politics has traditionally focused on industrialized countries.[68] The systematic analysis of the role of scientists in political life throughout the industrializing world is nascent at best.[69] This chapter offers, on the basis of the contrasting experience of two large industrializing countries, a more generic perspective on the conditions shaping state-scientists relations across political-economic systems.[70]

First, levels of macropolitical consensus and bureaucratic autonomy can explain why different patterns of state-scientists relations emerge across industrial and scientific sectors. Thus, even in a field often associated with priority governmental support such as nuclear energy, scientific research in Brazil was neglected relative to allocations for the purchase of machinery, technology, and foreign training.[71] Argentina's CNEA sheltered scientific research, and its scientists managed to "ride the coattails" of an industrial program privileged with budgetary and programmatic autonomy. Hirschman (1971) differentiates between "privileged" problems, where victims have adequate access to policymakers who are obliged to listen for the sake of political stability and their own survival, and "neglected" problems, where such direct access is missing. The role of scientists in forging such causal links between "neglected" and "privileged" issues (scientific development versus nuclear energy, in the case of Brazil) has been critical under exclusionary regimes where channels of access were blocked for most other sectors.

Second, the episode of collective action by Brazilian scientists analyzed in this chapter challenges two views: (1) the characterization of developing societies (by modernization theories) as those in which "there is not enough of a culture of the scientific community to stand up against the tradition of the civil service and the distrustfulness of politicians" (Dedijer 1968:159); and (2) the claim that scientists in the developing world refrain from taking an active role through their professional organizations in issues of atomic energy.[72] Neither an internal professional structure assumed to be geared only to the regulation of professional advancement nor the external barriers imposed by a politically closed and exclusionary decision-making context prevented professional associations and the SBPC from organizing the political opposition to the nuclear program. The internal cohesiveness, institutional strength, and public recognition of leading scientists were important in consolidating the group's impact on nuclear policy. The existence of different conditions in countries like Iraq and Iran, perhaps related to an even more brutal grip over civil society, may explain the absence of collective action by their sizable scientific communities. In the case of Iraq, however, the regime's sponsorship of an ambitious integrative scientific and technological effort (in this case more clearly geared to military objectives) reminiscent of Argentina's CNEA may well explain how the large-scale co-optation of scientists can neutralize conflicting political proclivities.[73]

Third, the experience of Brazilian scientists in nuclear policy defies some generic contentions (as opposed to the preceding specific conditions of industrializing countries) about the behavior of scientific communities vis-à-vis their main patron, the state. It specifically challenges the view that the scientific leadership has tended, "almost without exception, to acquiesce in any fundamental confrontation with the state, especially when opposition was likely to evoke serious sanctions."[74] It also weakens the claim that scientists mobilize merely to defend the autonomy of science from the intrusion of political forces.[75] In fact, the state's very neglect of the socioeconomic relevance of the endeavors of Brazilian scientists triggered the scientists' political activism. Furthermore, on the debate over the sources of scientific inquiry, the study suggests that the positions advanced by Brazilian scientists in nuclear policy reflected their tendency to subsume the scientific and technological system under a broader political-economic agenda. Thus, external determinants (political independence, social cost considerations) entered the conceptual phrasing of scientific problems, as in the case of the thorium group and energy research.[76]

Fourth, the study reinforces the historical predominance of physicists—particularly during the second half of the twentieth century—in encounters between scientists and the state, across different systems.[77] Internal characteristics of the discipline,[78] its centrality in the modern world, its heavy dependence on state support relative to most other sciences (because of the high costs of research and the open-ended value of research outputs),[79] have all been used to explain that predominance. Brazilian physicists were, in fact, better organized in their public indictment of the regime than any of their colleagues in other scientific disciplines, particularly after the dissolution of the thorium group and the regime's attacks on the University of São Paulo and the Rio Center for Physics Research.[80] As in other cases—notably the former Soviet Union and the People's Republic of China—physicists also have tended to be singled out for persecution, a phenomenon largely validated by this chapter's overview of Argentina and Brazil.[81]

As to the political-ideological content of the physicists' activism, it is important to note, first, that most of the physicists criticized Brazil's agreement with KWU while arguing for an alternative nuclear path, stressing "national independence" and domestic scientific spin-offs. This was, for the most part, no antinuclear movement, but one emphasizing the protection of local scientific-technological resources, not an unfamiliar theme in the experience of western European, Japanese, and other scientific-technical communities. Second, and despite a primary focus on the political-economic nature of their nuclear programs, Argentine and Brazilian physicists—like many among their U.S. and Soviet counterparts—were prominent in the opposition to the military uses of nuclear energy and mobilized their communities in support of the creation of a nuclear-weapons-free zone in the Southern Cone.[82] At the same time, many challenged the legitimacy of the Nuclear Nonproliferation Treaty, the London Suppliers Group, and all perceived attempts by the nuclear powers to freeze the existing pattern of international nuclear stratification.[83]

Finally, this chapter is threaded with evidence of schizophrenic state behavior vis-à-vis scientific development, particularly at the basic end of the research spectrum. This contempt for science, coupled with uncertainty about its potential contributions, has characterized many twentieth-century states.[84] The tension was certainly not resolved in the Argentine context. The ambiguity of the Brazilian regime was evident in the pattern of following scientists' purges with attempts to rebuild a research infrastructure. Persecutions and inducements to emigrate alternated with efforts to reverse the brain drain.[85] Policies limiting participation of scientists in international meetings were intertwined with attempts to give the scientific community international visibility. This ambiguity may not be completely resolved until the historiography of technical progress settles the controversy over the actual extent to which technological development depends on science. There is significant agreement, however, that such dependence has increased substantially over the past century.[86] The emphasis on building a strong domestic *technological* infrastructure has gained, of course, widespread recognition, although there has been disagreement on strategies to bring it about. It is to this particular angle of the Brazilian and Argentine nuclear industries that I turn next.

PART THREE

The Outcome

CHAPTER 6

*Public and Private
Technological Gains:
The Direct Effects of Nuclear
Industrial Investments*

We are now familiar with the political-economic sources of
the Brazilian and Argentine nuclear industries. Part III relates the
policy and bargaining processes analyzed in Parts I and II to outcomes, particularly regarding technological development. The literature on technology transfer has paid relatively little attention to a
most important set of questions regarding "how well the transferred
technology is absorbed, diffused and built upon in the host country."[1]
This chapter examines the direct technological effects of the inception of nuclear industries in both countries. It begins with an analysis of the state as entrepreneur, including an evaluation of efforts in
R&D in the nuclear sector. The next section surveys the performance of state agencies in their regulatory capacity, particularly in
screening technology transfers and in strengthening the links among
state, science and technology, and industrial activities. A third section explores the microeconomic angle, summing up the technological gains of private firms involved in the nuclear program. The conclusions highlight some of the general implications of my discussion
for the role of the state in technological development and for the nature of learning and technological change from a micro, firm-level
perspective.

The State as Entrepreneur

TECHNOLOGY TRANSFER AND
ABSORPTION AT NUCLEBRÁS

Nuclebrás was the institutional reflection of a nuclear policy compatible with the regime's consensus over Brazil's economic model. Such policy implied heavy reliance on state entrepreneurship and on easily available foreign technology. Created in 1974, Nuclebrás was a logical extension of a policy of state expansion into "empty spaces" of the productive structure, in many cases through an alliance between public sector technicians and the military.[2] Even prior to the 1960s, the public sector had traditionally played a key role in assuming responsibility for bottlenecks—such as building electrical generating capacity—that often attracted neither local nor foreign capital because of both their low returns relative to other industries and the higher risk of expropriation.

As suggested in Chapter 5, critics of the nuclear program regarded the joint ventures between Nuclebrás and German firms as a classic model of "aborted" technological development. In their view, the acquisition of foreign technology reflected a narrow and immediate cost-effectiveness criteria that ignored opportunity costs and learning-by-doing.[3] Imported technology implied the exclusion of domestic research institutions and their condemnation to a marginal role in the country's industrialization. This exclusion arguably imposed additional costs, such as the inability to adapt the technology to local needs and resources. A prime target of critique was NUCLEN, the joint venture (75 percent Nuclebrás, 25 percent KWU) established in 1976 as Nuclebrás's subsidiary responsible for design architecture/engineering. NUCLEN's functions included power plant procurement and follow-up, erection, and construction management;[4] it was also to serve as the interface for transferring technology from German firms to domestic engineering firms. As the agent entrusted with "whitening the black box" (or disaggregating the technological package), NUCLEN had several different options, listed below according to progressive levels of technological effort. It could: (1) manage the acquisition of a black box with little local technological input; (2) begin to unpack the project through a detailed analysis of supplies and costs, with a view to increasing local sourcing of subsystems;[5] (3) perform the analysis of specifications, development of equipment, technical assistance to suppliers, assembly, quality control, and licensing; (4) assume the project's management, thus increasing its leverage over technological decisions;[6] and (5) carry out important changes and innovations in conception and design.[7]

NUCLEN was initially conceived to fulfill the second and third functions by establishing a Department of Industrial Promotion responsible for supplier development and planning progressively higher indices of national participation in conventional (nonnuclear) components from 30 percent for the first two plants (Angra 2 and 3) to 70 percent for the last (eighth) plant.[8] For the first two plants, NUCLEN was to be responsible for civil, piping, and plant layout (interferences), and for providing 50 percent of the engineering required, including 91 out of 113 mechanical systems.[9] NUCLEN actually supplied 105 of these systems, including portions of the nuclear steam supply system (see Appendix D). All mechanical systems for the planned third and fourth plants were to be NUCLEN's responsibility.[10] Mechanical systems, however, could be considered peripheral—in technological terms—to the reactor systems, and effective absorption of the latter systems would begin only with the fifth plant onwards.[11] Until then, KWU was to be responsible for the design of reactor systems. Owing to the slowdown in the program, KWU later agreed to subcontract part of the services for the reactor systems with NUCLEN, departing from the original schedule. Accordingly, NUCLEN took over all nuclear auxiliary systems and acted as subcontractor for KWU in areas related to reactor systems.[12] Project responsibility remained in KWU's hands for the first four plants, and, had they been constructed, NUCLEN would have taken over that responsibility for the last four.

Critics held that NUCLEN's technological absorption was in fact limited to producing identical copies of German plants. Moreover, they argued, KWU protected the most important technical aspects very effectively, placing them beyond NUCLEN's reach.[13] Former NUCLEN director Joaquim Carvalho, for instance, argued that performing the detailed design of the basic KWU project would not enable NUCLEN to introduce important changes in the original design.[14] This was particularly the case, he argued, because technology could not be absorbed without intense interaction between industrial research centers, project designers, constructors, and compo-

nents producers.[15] According to Carvalho, NUCLEN's cooperation
with private Brazilian engineering firms and research centers—such
as the Institute for Energy and Nuclear Research (IPEN), the Institute
for Technological Research (IPT) in São Paulo, and the Institute of
Nuclear Engineering (IEN) in Rio de Janeiro—not only would have
resulted in lower costs, but also would have improved effective technology absorption.[16] Well-trained technicians would have been better
equipped to interact with KWU engineers than the young, inexperi-
enced technicians NUCLEN itself trained.

In response to these criticisms, advocates of the nuclear program
argued that Nuclebrás was to integrate, under state leadership, all
industrial activities related to fuel-cycle and power reactor design
and construction, through a process of staged technology transfer.
They pointed to measures of successful technological absorption,
such as NUCLEN's performance of the complete seismic analysis for
the second plant, Angra 3 (the first was carried out by KWU), its development of most electrical, and all alarm and communication, systems, as well as all instrumentation and control for those mechanical
systems under NUCLEN's responsibility. As a measure of domestic
engineering participation, NUCLEN and private Brazilian firms were
to perform three-fourths of the required person-hours for the first two
plants, leaving one-fourth to KWU and associates.[17] NUCLEN was
also involved in adaptations of the basic KWU design to local conditions.[18] Such adaptations (referred to as "tropicalization") were
mainly in areas related to ocean proximity, redesign of cooling water
systems, and the conforming of the original computerized techniques (written for Siemens computers) to IBM systems partly produced in Brazil. According to NUCLEN's Superintendent of Mechanical Systems, Dr. Witold Lepecki, the plants to be designed by
NUCLEN would not be mere copies of KWU's reference plants.[19]
NUCLEN could use the German design to devise a smaller plant
(with three, rather than four, loops in the water system). Moreover,
the KWU experience would facilitate NUCLEN's potential absorption of breeder technology. In the end, NUCLEN officials argued, effective technological absorption and learning-by-doing required the
completion of the whole program (eight plants). That never happened, as we know, because only the first plant (Angra 2) was spared
from dramatic revisions, in the 1980s and 1990s, of the 1975
agreements.

In sum, both sides could marshal some evidence to support their fundamental positions. Critics could argue that the development of engineering capabilities at NUCLEN was limited to a transfer of experience in project coordination and assembly of nuclear plants, which did not include mastering of basic engineering and conceptual design skills. Nuclebrás officials could point to adaptations of the basic project in structural areas (such as support of the reactor vessel, layout, and others) and to input substitution, as evidence of technological gains from the nuclear program.[20] Similar debates raged over the technological performance of NUCLEP (heavy components), established as a joint venture between Nuclebrás (75 percent) and KWU, GHH, and Voest (25 percent).[21] NUCLEP was responsible for the design, development, and manufacture of heavy nuclear components for the nuclear steam supply system, the turbogenerator system, and complementary systems.[22] As discussed in Chapter 3, NUCLEP was eventually dismantled and its responsibilities transferred to private Brazilian firms in response to pressures for privatization.

The technological performance of the joint ventures in nuclear fuel-cycle activities created by the 1975 agreements (see Appendix B) was found to be seriously flawed. Instead, Brazilian agencies in fuel-cycle activities not related to those agreements wielded significant achievements. NUCLAM (51 percent Nuclebrás, 49 percent Urangesellschaft) was to be responsible for 10 percent of Nuclebrás's efforts in uranium prospecting, research, development, mining, and production of concentrate.[23] A decade after signing the agreements (1985), NUCLAM was not yet operational, while Nuclebrás's independent mining activities placed Brazil's uranium reserves in fifth place worldwide.[24] Also outside the agreements with Germany was Nuclebrás's Industrial Complex at Poços de Caldas, including a mining and a chemical treatment plant. Poços—the first industrial producer of uranium concentrate (yellowcake) in Brazil—was inaugurated in 1982, only two years behind schedule.[25] In the area of fuel-element fabrication, the German firm RBU (a KWU subsidiary) provided the Resende facility with the plant's basic design, while Brazilian firms (Hidroservice and Rodrigues Lima) were responsible for detailed engineering and construction, respectively.[26] The plant restricted itself to the loading, welding, and final assembly of UO_2 pellets, because RBU supplied the structural parts and uranium pellets.[27] Inaugurated in 1982, the plant had no orders because of delays in power plant construction.

In earlier chapters I discussed the criticisms raised against the joint venture NUSTEP (50 percent Nuclebrás, 50 percent Steag), located in the Federal Republic of Germany and holding the patent to the (jet-nozzle/Becker) enrichment procedure. Central to these criticisms was Steag's apparent unwillingness to transfer technology to Brazil. Another joint venture in uranium enrichment—NUCLEI (Nuclebrás Enriquecimento Isotopico S.A., 75 percent Nuclebrás, 15 percent Steag, 10 percent Interatom)—was to construct and operate a demonstration facility using the jet-nozzle process.[28] Impatience with the slow progress in the development of this technology led to the cancellation of these joint ventures in 1986. By 1987, unrelated national efforts yielded success in uranium enrichment through ultracentrifuge techniques (the same technology pursued by Saddam Hussein before the dismantling of Iraq's nuclear program in the aftermath of the Gulf War). Finally, original stipulations for a reprocessing plant with technical assistance from the German consortium KEWA/UHDE were discontinued in the early 1980s, while a national reprocessing project thrived.[29]

Most efforts to develop domestic technologies in the fuel cycle accelerated in the early 1980s, outside the 1975 agreements and under CNEN's renewed lateral autonomy.[30] The Commission cooperated with universities, research institutes, and private firms to develop agricultural and medicinal products, antipollutants, and equipment for nuclear reactors.[31] This new approach to technological development, under the new macropolitical and institutional conditions of the 1980s, bears strong resemblance to the model embraced by Argentina's CNEA from its inception.

ARGENTINA'S CNEA: UNPACKAGING AND TECHNOLOGICAL PERFORMANCE

The Argentine nuclear program had a longer history than that of Brazil. CNEA's early experience in the design, engineering, and construction of its research reactors enhanced its professional capacity for later decisions involving feasibility studies and the construction of a first power reactor in 1968.[32] In the sheltered environment of the Navy's fiefdom, CNEA officials developed, at the outset, a policy of maximizing the use of domestic technological resources. The means to that end, in the area of power plant engineering and components, were embedded in the nature of contractual instruments favored by CNEA negotiators. Unlike Brazil's network of joint ventures, CNEA purchased its first two power plants (Atucha 1 and Embalse) as turn-keys. In their unpackaging efforts, both Brazil and Argentina had acquired "grey boxes"; in other words, the plants were neither black boxes nor self-reliant local designs. However, while Brazil confined itself primarily to analysis of supplies (stage two) and, only to a limited extent, to the development of local suppliers (stage three), Argentina's efforts in stage three were heroic; in fact, CNEA went beyond that stage, to acquire several components locally and to seek project management (stage four). Argentina's entry into stage four in 1977, when CNEA had to assume direct responsibility for the second plant (Embalse), was a consequence of a disagreement with Atomic Energy of Canada Ltd., which requested Argentine compliance with safeguards in exchange for technology transfer. Rejecting that linkage, CNEA promoted the creation of engineering consortia to ensure continuation of the project. Thus, paradoxically, a significant leap in technological growth in Argentina occurred as a result not of a contractual stipulation with the supplier but an alleged contractual departure by the supplier.

For Argentina's third plant (Atucha 2), CNEA created ENACE S.A. (Argentine Nuclear Electrical Power Stations Company Limited) in 1980. This joint venture, similar in ownership structure to Brazil's NUCLEN (75 percent CNEA, 25 percent KWU), became the industrial-architect for Atucha 2 (scheduled to operate by 1997). There is little coincidence in the fact, at least from the perspective adopted in this book, that this new technological instrument—a departure from CNEA's earlier tradition—came into being in the wake of changes in macropolitical consensus and CNEA's lateral autonomy (as described in Chapter 2). These changes implied greater tolerance for some special, restrictive clauses (often associated with bargaining asymmetries in technology transfer) similar, to some extent, to those we encountered in Brazil's 1975 agreements.[33] KWU gained considerable control, including preference in the supply of turbogenerators and other conventional components, with some leeway by CNEA in price control. However, KWU's share was to diminish gradually and disappear with the fourth plant, no later than in 1995.[34] In fact, ENACE itself was conceived of as a temporary arrangement to be replaced by local engineering firms from the private sector, revealing the strength acquired by those firms as a consequence of CNEA's early nurturing of national entrepreneurs.[35] The outcome of earlier CNEA policies—a strengthened local private sector and an involved technical expertise—was now shaping new technology transfer arrangements. The most important contracts for Atucha 2 were signed immediately after the creation of ENACE, and criteria not completely favored by KWU did prevail in at least some of those contracts.[36]

In sum, both Brazil and Argentina acquired significant experience in reactor design and manufacturing through diverging contractual mechanisms. By 1987 CNEA could wield one locally designed pressurized heavy water reactor prototype and another (Argos) designed by ENACE as the products of its effective absorption of power plant engineering. Brazil designed a medium-sized power plant (100 MW) by the early 1990s, in an effort unrelated to the program implemented by Nuclebrás.[37]

In the area of fuel-cycle technology, Argentina developed techniques and equipment for airborne uranium prospecting and for a controlled operation to produce uranium concentrate.[38] CNEA designed and built a conventional uranium ore processing plant at Malargüe, in operation since 1965. Conversion from yellowcake to uranium dioxide took place in Germany until 1983, when a local facility was completed.[39] As early as 1957 CNEA decided to develop a fuel-element manufacturing capability, carrying out intensive R&D in fuel-element production and introducing important improvements at very early stages.[40] CNEA developed a completely new process for production of U-Al alloy (later patented in Argentina, the United States, West Germany, and Japan). For the commercial fuel-element production plant inaugurated in 1982, CNEA developed the plant's basic engineering in cooperation with the Karlsruhe Nuclear Center, performed detailed engineering on its own, and oversaw construction and assembly by a joint Argentine-German consortium.[41]

Research and development in the area of metallurgy allowed CNEA eventually to develop basic engineering for fabrication of fuel-element cladding tubes. There was progressive backwards integration (production of zirconium sponge) that allowed Argentina to place zircalloy tubes high on its lists of nuclear exports.[42] Enriched uranium for research reactors had to be imported from the United States until Argentina announced, in November 1983, that it had mastered the technology of uranium enrichment, a technical achievement shared by only a few other nuclear power states.[43] CNEA also operated a small experimental reprocessing facility separating plutonium on a laboratory scale by the late 1960s and began construction of a commercial reprocessing plant.[44] Finally, Argentina's first two power plants depended on U.S. and Canadian heavy water. Thus in 1980 CNEA purchased a turnkey heavy water industrial plant (Arroyito) from the Swiss firm Sulzer Brothers and began construction of a parallel pilot plant with its own resources. The industrial plant began operating in 1993, many years behind schedule.

EVALUATING RESEARCH AND DEVELOPMENT EFFORTS

Perhaps the most striking contrast in the technological performance of the two programs was in R&D efforts and investments. The following comparison bears in mind the institutional differences between the two nuclear sectors emphasized in earlier chapters. Whereas in Brazil Nuclebrás was the industrial agent and CNEN (the Nuclear Energy Commission) the licensing and regulatory organism, in Argentina CNEA (the Atomic Energy Commission) combined all these functions. The role of R&D within Nuclebrás was marginal indeed, but many of these functions were performed by CNEN, and increasingly so since the early 1980s. We can thus compare the programs' commitments to R&D through four major indicators: emphasis on highly qualified research personnel, actual budgetary allocations for R&D, degree of coordination between R&D efforts and the industrial program, and the ratio of domestic to imported research efforts.

The development of competent domestic technical personnel is a critical component in technology absorption. Defining the composition of this technical corps can, therefore, be considered a measure of technological effort. In December 1974, while negotiating with KWU (for plants two through eight) Brazil's CBTN (later Nuclebrás) had a total of 1,603 employees. The Division of Technology and Development had about 600 people overall.[45] Argentina's CNEA had at least 2,500 scientists, engineers, and technicians when it negotiated the supply of its first power plant and 4,000 for the second plant.[46]

In Brazil, Nuclebrás's R&D arm—the Center for Development of Nuclear Technology (CDTN)—began operating only in 1978, three years after the 1975 agreements on technology transfer. CDTN was designed to provide R&D services to Nuclebrás's subsidiaries and to train personnel to ensure successful absorption of technology and development of local know-how. As described in Chapter 5, leaders in the academic community regarded the nuclear program's allocations to R&D as quite marginal, pointing to CDTN as an agency involving only a few (technically limited) projects approved by KWU and unable to attract serious scientists.[47] CDTN included an academic work force of 210 people (one-third scientists and two-thirds engineers). Of these, 5 percent held Ph.D.'s, 30 percent held M.Sc.'s, and 25 percent were graduates in basic nuclear technology.[48] If one includes CNEN's personnel, the total number of academically trained researchers reached over 500.[49] The number of high-level professionals at CNEA, in all its departments, reached over 1,100. About 500 physicists, chemists, and engineers and 500 specialized technicians worked for CNEA's R&D department only.[50]

Research and development budgetary allocations are another important measure of technological effort. CNEA devoted an average of 10 percent of its total budget for R&D (about $150 million).[51] Total R&D allocations in Brazil included both CNEN and CDTN. CDTN received between $5 million in 1978 and $14 million in 1982.[52] CNEN devoted about 10 percent of its budget to R&D—$3 million— in 1980.[53] Combining CNEN's and CDTN's figures provides an approximation ($17 million) of Brazil's very small annual R&D investment, relative to that of Argentina.[54]

Beyond these figures, the degree of coordination of R&D activities with the nuclear industrial programs in progress is another important measure by which to gauge technological effort. Research at CNEA's three centers (Bariloche, Ezeiza, and Constituyentes) and its headquarters ranged from highly theoretical to applied areas, in an attempt to strengthen the interaction between technology-oriented scientists and scientifically competent technologists. CNEA's Department of Technology was a tangible example of this successful merging of science and technology. It was here where the "metallurgy group" (Grupo de Metalurgia) of engineers, physicists, and chemists absorbed graduates of CNEA's Institute of Physics in Bariloche.[55] This group led to the achievements in fuel-element technologies mentioned previously. In contrast, Brazil's CDTN emerged only after unrelenting criticism of Nuclebrás's poor R&D efforts. Using some of the Institute of Radioactive Research's (IPR) old infrastructure, CDTN revived previous research projects on development of a mixed-fuel (uranium and thorium) cycle,[56] provided technical support for NUCLEI's first cascade in the uranium enrichment demonstration plant,[57] and began developing software for the complex operation of nuclear plants.[58] CDTN's contribution to Brazil's industrial program was far more limited than CNEA's R&D arm.

Finally, an important measure of commitment to domestic R&D is given by the ratio between expenditures in this area at home and the cost of imported technology. As argued in Chapter 4, Brazil's 3:1 ratio of imported-to-national R&D expenditures contrasted with Argentina's reversed record, which approximated 10:1 domestic-to-foreign allocations. Argentina's investments in R&D yielded 43 patents, of which CNEA was responsible for 32 (related to materials and structural and equipment engineering). Argentine firms accounted for 23 percent of the patents approved in the nuclear area.[59]

The State as Regulator

SCREENING TECHNOLOGY TRANSFER

The benefits of state intervention in regulating technology transfer became a focus of scholarly attention and governmental interest worldwide beginning in the 1960s and particularly in the wake of the Japanese "miracle."[60] Brazil created the National Institute of Industrial Property (INPI) in 1970 to screen technology transfer agreements, including licensing for exploitation of patents, licensing for the use of trademarks, product and process technology (for consumer and capital goods), and engineering services. Consultations with INPI for technology transfer agreements became compulsory only in 1975, the year the nuclear "umbrella" agreements were signed. INPI began prohibiting restrictive clauses (such as preventing recipient firms from exporting products manufactured with the acquired technology) and demanded full disclosure from technology owners and a recipient's program of technology absorption.[61] Implementation of these measures was problematic, because while INPI oversaw the importation of know-how, it did not regulate the importation of equipment or materials; this was the domain of CACEX, the foreign trade bureau at the Central Bank.[62] Lack of effective coordination regarding these procedures, particularly where they faced incompatible macropolitical constraints, contributed to a faulty and selective implementation.

Most of these problems were evident in the nuclear agreements.[63] Over 30 licensing agreements signed between Brazilian and German firms were transferred to INPI for registration, to ensure the unavailability of national equipment of similar specifications. The agreements included transfer of basic designs, innovations by the technology supplier, and improvements by the recipient. The agreements also reserved the supplier's right to approve the recipient's detailed design. Moreover, the recipient firm was to refrain from exporting either the technology or the equipment produced with it; this restriction was not formalized—INPI prohibited such clauses—but sanctioned by a "gentlemen's agreement."[64] The special clauses and technical committees at Nuclebrás's subsidiaries also overruled, in many instances, formal legislation regarding technology transfers.[65] In other instances, CACEX's responsibility for verification of the unavailability of similar national equipment was transferred to the Industrial Development Council (CDI), which, in turn, granted NUCLEI exemptions from import taxes.[66] Thus, the performance of agencies entrusted with screening technology transfer agreements (in terms of costs, restrictive clauses, and preference for "real" technology rather than imported capital goods) suggests that their prerogatives were compromised, as in other industrial areas, when they trampled with core macropolitical priorities.

INPI's intervention in technology transfer screening intensified by the early 1980s, relying by this time on consultations with domestic research institutions to survey locally available technology and resources.[67] The New Normative Act No. 15 of 1982 prevented state enterprises from imposing foreign technology contracts on their local suppliers and allowed foreign services only when these were not available at home. Although the 1975 agreements were an irreversible reality by that time, INPI initiated cooperation with the Brazilian Nuclear Quality Institute (IBQN) in the evaluation of technology transfer agreements.[68] Once again, these shifts can be traced to the erosion of the structural and institutional context within which the original agreements came to life.

INPI's counterpart in Argentina was created only in 1971, several years *after* the first nuclear contracts were signed.[69] It did not play an important role in the screening process in the nuclear sector, even in subsequent contracts (in fact Argentina's INPI was frequently criticized for many of the same weaknesses that affected its Brazilian counterpart). However, CNEA's own screening capacity preceded any central government efforts to regulate incoming technology, and CNEA's lateral autonomy helped it maintain a monopoly over that function, even after centralized efforts were in place.

STRENGTHENING THE "SÁBATO TRIANGLE"

Creating and nurturing synergistic links among science, industry, and the state is often considered a major requirement for scientific and technological development. In the Latin American literature, these links are frequently referred to as the "Sábato triangle."[70] Article 1 of the 1975 agreement with West Germany and KWU established the need for cooperation among Nuclebrás, Brazilian research institutions, and the productive sector in all industrial phases of the nuclear program.[71] The outcome, however, suggests otherwise.

Integrating national research institutions with industrial activities in the nuclear sector was a pillar of the "Sábato triangle." Yet, Brazil's CNEN—responsible for licensing, regulation, safety training, and research—played a somewhat marginal role in the implementation of the 1975 agreements, at least with respect to the last two functions. The segmented nature of nuclear decision making, and CNEN's subordinate role (to broader energy and industrial requirements), prevented it from coordinating effectively among university, academic research centers, and nuclear industrial activities.[72] Nuclebrás was even less interested in national nuclear research activities, as previous sections clearly suggest. In fact, Nuclebrás's director, Paulo Nogueira Batista, opposed funding domestic projects that could divert funds from the agreements with Germany.[73] IPEN, the Institute for Energy and Nuclear Research, was bypassed, despite its twenty years' of experience in fuel-cycle technologies.[74] An agreement signed with Germany in 1978 for research on thorium cycles excluded Brazil's leading experts on the subject, members of the disbanded "thorium group" of Belo Horizonte, discussed in Chapter 5.[75] Nuclebrás rejected former NUCLEN official Joaquim Carvalho's efforts to consult with national academic institutions (such as the University of São Carlos's Department of Material Sciences, known for hosting high-level metallurgists) or with other state agencies (the aeronautical firm Embraer, with vast experience in casting and other techniques).

Coordination in the area of training was one of the weakest links in the program at the outset. Heeding to sharp critics, Nuclebrás helped establish Pronuclear (Program of Human Resources in the Nuclear Sector), an interministerial program headed by the presidents of CNEN, Nuclebrás, the National Research Council (CNPq), and representatives of the Ministry of Mines and Energy and the National Security Council. Pronuclear granted a total of 403 fellowships to five academic institutions between 1977 and 1981.[76] Over 384 engineers, physicists, and chemists participated in these specialization courses in nuclear technology; some found employment in Nuclebrás and subsidiaries.[77] This record was not sufficient to counter the claim that, at least until the early 1980s, Pronuclear failed to coordinate the industrial aspects of the nuclear program with training in academic centers.[78]

Macropolitical and institutional changes in the 1980s brought about improved coordination between the nuclear industry and academic centers.[79] Recognizing the weak links between the two, the Third Basic Plan for Scientific and Technological Development (1980–85) recommended ways to strengthen them and proposed the permanent allocation of a percentage of investments in the nuclear industry for R&D activities. CNEN, now endowed with greater lateral autonomy, began increasing its cooperation with academic research in 1982.[80] Its new pivotal role in integrating industrial and research activities now resembled CNEA's model as well as CNEA's tradition of bridging scholarship and applied tasks. Very early on, CNEA signed an agreement with the University of Cuyo whereby physics and nuclear engineering were to be taught at CNEA's Balseiro Institute in Bariloche.[81] These efforts were consolidated in the stable matrix of relationships between the Balseiro Institute and CNEA's Department of Technology.

A second major pillar of the "Sábato triangle" approach to technological development emphasized state responsibility in stimulating the development of technological capabilities at the level of the firm. The presumed reluctance of private firms in developing countries to invest in technological development brought the importance of such intervention to the fore. As Erber (1984) suggested, the significant differences between the private—as opposed to the social—calculus of gains and costs of innovation in capital goods require state intervention for promotion of capital goods. After all, even in countries with strong entrepreneurial sectors such as West Germany and Japan, the state underwrote the costs of R&D in the nuclear industry. The external environment can influence technological change within the firm through incentives and penalties (by encouraging firms to improve their technology and to develop technological resources), through networks of specialized agencies facilitating the flow of technological information, and through programs for vocational and in-plant training.[82]

Brazil's National Economic Development Bank (BNDES, or BNDE until 1982) became the most important agent for the development of the national private sector in the 1960s and 1970s. It designed its special fund for science and technology (FUNTEC) in 1964 to support development of human resources in scientific and technological disciplines, to stimulate and support research and innovation by private national firms, and to promote adaptation of foreign technology to local conditions.[83] In 1974, FUNTEC's emphasis shifted from supporting universities and research institutes to strengthening demand and utilization of local technology in industry.[84] This trend was reinforced by a series of measures to support private firms, through coordination of state procurement, fiscal, and financial incentives.[85] BNDES devoted an average of 25 percent of its funds to the energy sector during the 1970s. Yet nuclear energy never accounted for more than 1 percent of all funds for energy projects. The Bank's annual report justified its own marginal participation in the nuclear program by pointing to the budgetary autonomy of the nuclear sector.[86] In reality, BNDES experts questioned the viability of the joint venture (state-foreign) NUCLEP, particularly the state's incursion into areas where private firms could allegedly have participated effectively.[87] Notwithstanding such skepticism, Brazilian firms producing for the nuclear sector received some BNDES support—in no timely fashion—through one of its subsidiaries, FINAME, mostly to substitute imported materials and parts for local ones.[88] All such attempts at substitution required approval by KWU and testing by the quality assurance agency IBQN. FINAME also provided financing to NUCLEP for purchase of machinery and equipment for the heavy components factory, to FURNAS (the utility/client) for financing of equipment to be produced in Brazil, and to Nuclebrás's CDTN.[89]

Another Brazilian agency—FINEP (Financial Agency for Studies and Projects)—expanded its functions in 1971 from merely financing feasibility studies to channeling the resources of the National Fund for the Development of Science and Technology (FNDCT) and appraising projects submitted to it.[90] Since 1973 FINEP's Program of Support of Technological Development of the National Enterprise was designed to finance the whole range of scientific and technological activities in a firm. Capital goods industries became the main recipients, 80 percent of them in relation to product development.[91] FINEP reportedly provided the utility FURNAS with some financing for nuclear equipment to be produced in Brazil.[92] Its marginal support for private firms working in the nuclear sector was particularly evident in light of FINEP's significant support for alternative energy technologies.

Finally, Nuclebrás's joint venture in nuclear engineering, NUCLEN, was formally responsible for aiding private firms in the qualification process and for supporting development of quality assurance and quality control practices. NUCLEN was to act as a mediator, supervising licensing agreements between Brazilian firms qualified by IBQN (the Brazilian Nuclear Quality Institute) and German suppliers selected by NUCLEN. In reality some Brazilian firms were forced by NUCLEN into these agreements, despite their extant capacity in the technology in question.[93] Attempts by former NUCLEN official Joaquim Carvalho to stem KWU's pressures for licensing from German firms failed and led to his resignation in protest in 1979.[94] The conception of NUCLEN as a buffer between the supplier and the recipient does not appear to have had the intended effect of protecting private firms from abuses by suppliers or to have improved the process of technology absorption in a significant way.

In Argentina, awareness of barriers to be posed by central economic agencies forced CNEA—on the basis of its lateral autonomy—to rely on its own budgetary sources to finance private firms. Efforts to assist firms involved with the nuclear program began in 1962, when CNEA created the Services of Technical Assistance to Industry (SATI), *before* signing any commercial agreements for a power plant. SATI was set up in association with the Chamber of Metallurgical Industries to support Argentine industry; it placed its scientific and technical resources at the service of the country's industry.[95] Chapter 7 examines SATI's performance from the perspective of diffusing technological capabilities throughout industry. The next section discusses the agency's interface with private Argentine firms in the nuclear program.

The Microeconomic Angle: Technological Growth in the Private Sector

My analysis of outcomes has so far focused on state performance. Technological change at the level of the firm, however, is not only related to governmental promotion strategies, but is also a product of deliberate efforts within the firm. In this section I evaluate the technological performance of these firms and their implementation of technology transfer agreements and special clauses discussed in Chapter 4.[96] In Brazil, these firms are among the largest, oldest, and most sophisticated in their respective areas—the "crust" of the sector—big or medium-sized enterprises, with five hundred employees and over.[97] Domestic capital goods supplies came mainly from Brazilian firms dominant in their field and included mostly boilers and thermal equipment, and mechanical equipment and machinery.[98] These firms had prior experience with licensing from foreign suppliers, reliable machine tools and equipment, and advanced training programs.[99] Most were capable of designing and manufacturing complex capital goods, including "custom-built" components.[100] There is relatively little research in this sector, and Brazilian firms have a tradition of relying heavily on foreign designs.[101] All firms had practical experience in material adaptations, and four (out of five) had experience in process or product adaptation and improvements.

Production for the nuclear program reached 40 percent of total revenues for some of these firms (50 percent in a few cases). They invested in equipment, personnel, and quality control and quality assurance to prepare for nuclear supplies.[102] As argued above, Nuclebrás compelled them to sign technology transfer agreements with German firms for the production of specific components. The interviews I conducted confirmed that state agencies had applied technology transfer legislation in a lax fashion so that, for instance, explicit technology transfer costs (for components) ranged between 5 and 8 percent of the total cost, a higher level of royalties than that normally approved by INPI.[103] This flexibility, however, was not peculiar to nuclear agreements, and exceptions were common in other areas as well, particularly until the mid-1970s. There were also implicit costs embedded in requirements for imported materials and parts, and costs associated with technical assistance not included in royalties. In addition, there were instances of special clauses attached to the licensing agreements, such as the need for the supplier's approval of exports (as in the case of heat exchangers for nuclear plants).

In supplying the nuclear program, some of these firms improved their overall engineering capabilities in mechanical manufacturing, mainly through the adaptation of imported designs.[104] Adaptations also involved "tropicalization," which made products and production compatible with local climatic conditions and labor patterns. In some instances, however, the licenser itself performed the adaptations, in collaboration with local personnel. Most firms regarded these gains as limited indeed, arguing that German designs were not needed in many cases and were simply imposed on them by NUCLEN.[105] At least two firms felt their prior experience with Brazil's first reactor (Westinghouse's Angra 1) had enabled them to produce for Angra 2 without foreign assistance.[106] Although learning, if any, took place mostly in the area of production (not design), one firm acknowledged it acquired a capability for basic and conceptual design of heat exchangers for nuclear plants.[107] On balance, producers did not regard their own R&D capabilities (narrowly defined) as having been particularly strengthened by their involvement with the nuclear program. However, the nuclear program did strengthen the firms' R&D aptitudes in the broader sense of improving the manufacturing process itself.[108] Firms also acquired experience in handling structures of dimensions not yet produced in Brazil, such as the pressure vessel (500 tons) and heat exchangers (20 meters high).[109] Production of the reactor's containment wall, for instance, implied a high level of technical complexity in structural design, specified materials, and manufacturing process. All firms in the capital goods sector agreed that the nuclear program did not promote links with governmental and other research institutions, but rather reflected routine patterns of making very little use of research institutes by industry, mostly for testing and quality control.[110]

Argentine capital goods firms, particularly the largest ones involved with the nuclear program, shared a similar background with their Brazilian counterparts, in terms of age, investments in training, hiring patterns, performance in adaptations, reliance on foreign licensing and on their own general accumulated experience, as well as specific experience in steel mills, oil, chemical and petrochemical, and hydroelectrical industries.[111] While CNEA generally required foreign licensing, it emphasized agreements requiring genuine transfer of technology, including technical assistance and "on the job" training.[112] Even firms that routinely emphasized reliance on their own engineering department for troubleshooting had a strong tendency to resort to suppliers in the nuclear sector. Most basic designs were imported, although Argentine firms were able to carry out a small portion of the basic engineering of mechanical and electrical systems (for the conventional parts of the second plant).[113] CNEA also developed prototypes of components in instrumentation and control and supported local production of these items by small- and medium-sized firms.[114] A private firm supplied Argentina's third plant with heavy components (two steam generators, one pressurizer, and three moderator coolers) for the primary nuclear system; the state enterprise NUCLEP supplied such components in Brazil.[115] Argentine firms also produced the reactor's rolling bridge and the turbine's main condenser, which arguably required technological sophistication.[116] Several firms benefited from their participation in fuel-cycle facilities such as the heavy water plant.

CNEA's agency SATI aimed specifically at strengthening the links between state and private technological activities. The active participation of the Institute of Industrial Technology in metrology, testing, and related areas is a case in point. The gains from supplying the nuclear sector were particularly evident in improvements in the production process, including welding procedures, machining, finishing, testing, and other manufacturing and inspection techniques.[117] Of particular importance was the introduction of raw materials certification, previously nonexistent in Argentine industry. Firms established specialized quality assurance departments and purchased shop and quality control equipment. CNEA qualified about 120 firms to produce components for nuclear plants. Argentine private firms accounted for eleven patents of the 43 Argentina held in the nuclear sector.[118] For the fourth plant, private firms claimed they were able to supply over 60 percent of 225 systems previously imported.[119]

A brief overview of the engineering sector reveals the relatively high technical level of firms involved in services for the nuclear program. As with capital goods, Brazil's engineering concerns were large and well established;[120] they too relied heavily on foreign licensing or technical assistance agreements, and on foreign engineering firms and consultants.[121] The consortium Promon-Natron provided detailed design of mechanical systems (piping), while Promon-Engevix was responsible for civil structural design.[122] NUCLEN (KWU) provided core engineering services (conceptual and basic designs), leaving structural and piping engineering to private firms. Although these firms saw themselves fit to perform these tasks (including basic design), NUCLEN required the involvement of German consulting firms.[123] In other words, transfer of technology was considered to be limited to a transfer of German practices and experiences, norms and regulations, not always relevant to the materials, construction, and assembly practices of the recipient.[124] A more careful selection of the areas requiring foreign assistance (perhaps in the form of independent consultants) may have limited the costs and maximized the development of local design capabilities, preempting, in turn, the incorporation of unnecessary foreign codes, standards, and materials.[125]

Argentina's CNEA avoided these pitfalls only to some extent, although it managed to strengthen engineering firms significantly through assertive allocation of tasks. Some of these firms were far smaller prior to their involvement with nuclear supplies. The overall number of firms in the sector grew as well, and by 1985 about 22 engineering firms were participating in a bid for piping design for the third plant (Atucha 2). The most evident sign of the competence acquired in servicing the nuclear sector was the firms' eventual turn toward exports.[126] Producers of components turned outward as well, particularly in the area of pressure vessels and heat exchangers. Three firms participated in an Egyptian bid (involving components, engineering, and construction), in conjunction with a major international supplier.[127] All in all, nuclear power accounted for 43 percent of Argentina's total "project" exports (not including construction, licensing, technical services, direct foreign investment, or capital goods).[128] This figure should not obscure the important fact that, thus far, export profitability was doubtful, and prestige considerations overshadowed rational motivations, both for recipient and provider.[129] Moreover, there were tremendous opportunity costs to this focus on nuclear energy at the expense of a wide range of industrial sectors, where Argentina had long ago lost early advantages. In Chapter 7 I explore in greater detail the impact on their export performance of private firms' exposure to the nuclear program.

State Intervention and Technological Performance in Perspective

A review of technological performance in Brazil's nuclear program reinforces the claims made in earlier chapters regarding the macropolitical priorities that underpinned nuclear policy and their impact on Brazil's effectiveness in bargaining with technology suppliers. The military-technocratic regime that had fostered the rapid economic growth of the late 1960s to early 1970s sought the fast inception of a nuclear industry, in order to contribute to the immediate resolution of potential energy shortages that might have endangered the model's success. Such a plan required greater reliance on foreign technology than Brazil's nationalist rhetoric on scientific and technological development could bear. Thus, commitments to strengthen technological capabilities quickly became subordinated to the program's requirements. A relaxation of technology transfer regulations, delays in the coordination of financing and fiscal incentives, limited efforts to support the technological development of national private firms, and poor coordination with Brazil's research centers were all outcomes of these macropolitical constraints, which a segmented context of implementation only reinforced.

These conditions served technology suppliers well and allowed them to prevail in restricting technology transfer to the bare minimum required by the joint ventures, in forcing unnecessary licensing and restrictive clauses, and in undermining occasional efforts to integrate domestic research institutions. Thus, the agreements with KWU were far less rewarding technologically than sectoral activities unrelated to them. Foreign licensing imposed by NUCLEN (and KWU) suffocated, to some extent, local technological activities in the area of design. Paradoxically, Brazil took less advantage of the prior experience of its own capital goods and engineering firms, despite their greater propensity—relative to Argentine firms—to establish production facilities closer to international standards, to invest in R&D, and to export a proportion of their output.[130] Subsequent structural and institutional changes in the early 1980s brought technological activities in the nuclear industry in line with new sectoral objectives, at a time of low macropolitical consensus and heightened bureaucratic autonomy.

The technological performance of Argentina's CNEA reflected its early efforts to absorb scientific and technological clienteles into its fold. CNEA developed academic disciplines, professional training, and new academic and industrial institutions, introduced new techniques, processes, and materials, and aided the private sector technologically through SATI. It bore the price of technological development in the nuclear industry single-handedly—thanks to its bureaucratic autonomy and its staying power vis-à-vis Argentina's Army and Air Force—throughout decades of macropolitical turmoil. Although both programs were plagued by delays and lack of resources, these problems were much more pronounced in Brazil, where the level of "shielding" of the nuclear sector was never comparable to that of Argentina.[131] These factors impinged on the ability of Nuclebrás to recruit and maintain highly qualified technical personnel.

These different technological paths were reflected in the level of technological maturity reached by their respective nuclear sectors.

The advantages of Argentine firms became evident when Brazil and Argentina signed agreements to cooperate in the design and construction of nuclear plants.[132] Argentina was able to undertake project management responsibilities earlier, and led the way in power plant design and construction and in fuel-cycle capabilities (including exports), but not for long. Despite the observed deficiencies in its large-scale program with West Germany and its less directed efforts (relative to CNEA's), Nuclebrás gained considerable expertise.[133] From the point of view of state policies regarding technological development, this outcome suggests that the margin of advantages accumulated by a policy of protecting domestic scientific and technological resources may dissipate far more rapidly than the policy's advocates maintain.

At the same time, Brazilian firms' preexisting endowments in industrial technology had a lot to do with their ability to catch up. These endowments were the legacy of earlier state procurement policies, which these firms regarded as the most effective means of gaining technical expertise.[134] In fact, most firms qualified for nuclear supplies because of their prior production experience, design capability, and technical credentials.[135] Clearly, manufacturing know-how and experience with detailed design skills are preconditions for higher levels of technological competence. Although the nuclear program was far from a technological turning point, particularly for Brazil's engineering firms, the experience contributed generally to improving production techniques, quality assurance capabilities, and managerial capacity.[136]

These gains challenge, to some extent, classic assumptions of dependency theory associating licensing agreements with limited learning opportunities and with obstacles to the development of local technical capabilities.[137] These effects can be largely mediated by the firm's own efforts as well as governmental incentives. Argentine firms relied on foreign licensing after an effective technology search and a carefully negotiated technology transfer agreement (emphasizing learning-by-doing) in line with CNEA's requirement.[138] These firms' experience suggests that a concomitant investment in local R&D can reduce the disadvantages of licensing, particularly regarding technology absorption, obsolete technology, and restrictive practices.[139] Licensing did not prevent Brazilian firms from adapting and modifying the technology to suit local conditions or from contributing to technical training.

CNEA's own experience suggests that reliance on foreign technology does not invariably increase foreign equity or managerial control, and that it is possible to capture technological externalities even from black boxes (or gray boxes) by involving domestic entrepreneurial and technical resources.[140] This is especially the case when a single sectoral agency controlling the program's implementation considers technological learning as an essential *input*, rather than a potential *side effect*, of the introduction of a new technology. As with state enterprises in India and South Korea, CNEA tried to unbundle technology from foreign equity, searching for components (including machinery) separately, in a strategy geared to maximize the transfer of skills and the adaptation of foreign technology to local conditions.[141] In doing so, CNEA incurred no higher costs than did Brazil, despite the presumed lower expenses involved in straightforward technology transfers sensitive to market—rather than R&D—considerations only. The virtues of joint ventures proved to be less rewarding in technological terms than what Brazilian negotiators of the nuclear agreements had expected. Multinational corporations resist joint ventures that pose a threat to the competitive position of the parent company. Yet some elements in the motivational structure of the supplier can ensure greater willingness to supply information regardless of the structure of the agreement.[142]

Suppliers' efforts cannot prevent recipients' growing technological capabilities from competing with those of multinational corporations in the long term, in areas previously dominated by the technology supplier. The complete indigenization of nuclear power plants by industrializing countries does not seem to involve a significantly more extended process than that of the first-tier nuclear industries in developed countries. The role of public enterprises, across the import-substitution/export-oriented divide, has been more critical where private firms had lower levels of technological expertise. In all areas—including the fuel cycle—dedicated agencies have been able to combine domestic efforts with foreign technology to bear the expected fruits. The combined experience of the People's Republic of China, Taiwan, South Korea, Iraq, Brazil, and Argentina leaves little room for doubting the feasibility of diffusing nuclear technology internationally. What does remain questionable is whether efforts invested in this sector are worth their social costs, a question I undertake next.

CHAPTER 7

Myths and Facts:
Indirect Effects and Spin-offs
in Nuclear Industries

I usually compare Brazil's nuclear industry program — in developmental terms — with the Apollo Program for the United States.[1]

Será como un novo despertar en nossa indústria.
  (It will be like a new awakening in our industry.)[2]

Despite differences in conception and implementation, Brazil and Argentina shared with other industrializing countries a common assumption that the establishment of nuclear industries would propel their countries into higher industrial-technological stages.[3] That is, beyond its intrinsic contribution to the supply of energy, the nuclear industry was to help bring about a wider industrial transformation, through what are often referred to as spillover effects, linkages, external economies, benefits, spin-offs, or side effects.[4] No systematic study dedicated to examining this presumed relationship is yet available, and this chapter is no more than a first salvo in that direction. After some conceptual clarifications regarding the analysis of technological spin-offs, I explore their nature and extent in the Brazilian and Argentine nuclear programs from an aggregate perspective. I then examine these programs' effects on stimulating local entrepreneurship and on creating new industries—forward and backward linkages—with a special focus on exports.[5] I conclude by specifying methodological and conceptual limitations in the generic analysis of spin-offs, hoping to contribute to the renewed debate on spillover effects as a central concept in today's industrial and trade policy.[6]

Spin-offs: Problems of Definition and Operationalization

The concept of spin-offs implies that even hardware that was developed but is no longer extant (for instance, a particular reactor model) is a national resource, in that the knowledge acquired in its design and production can be applied to other products and processes to the benefit of industrial development.[7] The spin-off effect is even greater when the particular technology introduced embraces R&D within a large spectrum of scientific and technological disciplines. In this sense, spin-offs can bear some of the characteristics of public goods. It is impossible to compartmentalize the consequences of innovation, and one ought to look at interindustry effects.[8] In other words, because of the sectoral interdependence of technological change, technical skills and resources from one industry progressively spread to others. In the economics literature this horizontal transfer of technology within an industry is also defined as "diffusion" or "the means whereby innovations become part of the production function or product range of economic units which are not the originators."[9]

However, not only are these benefits not directly obscrvable, for the most part, but "the problem of appropriability makes it difficult to trace their distribution among the economic units involved."[10] Measuring the economic payoff to technological innovation is no easy task. As Rosenberg (1982) argues, the expansion of technological change from one industry to another defies easy summary or categorization. It is therefore hardly surprising that the literature on technological development among latecomers has overlooked the systematic study of interindustry spin-offs, beyond recognizing the importance of the synergistic effects of learning-by-doing, personnel movement, informal exchange of information, copying, and transferring.[11] As a first cut, one could rely on selected case studies (success stories of specific spin-offs) in evaluating these indirect effects. This exercise would have limited value, however, as evidence on which to draw broader generalizations. At any rate, even such modest efforts were not available when I began this analysis of spin-offs from Brazil's nuclear program, although a few studies addressed the Argentine case elliptically. An alternative path would be to simply list products or processes (learned and transferred); yet such an exercise clearly cannot provide a definitive measure of spin-offs, unless we are able to establish a priori, across sectors, what an appropriate rate of spin-offs is.[12]

With these caveats regarding the identification and measurement of spin-offs, this chapter surveys the major beneficiaries of technical changes introduced by the nuclear programs in Argentina and Brazil. Because few aspects easily render themselves to a quantitative analy-

TABLE 5

Nuclear Industry and Potential Spin-offs

Processes/skills	Materials/products	Potential industrial applications
MINING AND ORE PROCESSING		
Prospecting and exploration. Ore crushing, screening, grinding, washing, leaching, gravity separation.	High-quality portland cement; metal ores (zirconium, molybdenum, chromium)	Chemical Construction Optics/glass Metalworking
FUEL PROCESSING		
Chemical processing, gas handling, pelletization, ultracentrifuge, high-power laser, fuel reprocessing	Isotopically pure elements, fuel elements, radioisotopes	Chemical, petrochemical
MATERIALS AND METALLURGY		
Physical and mechanical properties (elasticity, fatigue, creep, plasticity, relaxation, corrosion resistance, phase transformation, diffusion, dislocation theory, alloy theory, electrical and thermal conductivities). Metal and ceramic production processes (heat treatment, extrusion, sintering, rolling, precision casting, foundry, forging, solidification, firing, refining of nonferrous metals). Finishing processes (surface, recrystalization, thermal/mechanical treatment). Nondestructive testing (helium leak tests, ultrasonics, radiography).	Ceramics. Metals (zirconium, molybdenum, etc.). Special alloys (specialty steels, zircalloy, etc.).	Chemical Construction Aerospace Metalmachining Transportation
MECHANICAL ENGINEERING		
Stress analysis; other destructive tests (compression, tension, shearing); machining processes (conventional machine tools, electron discharge, and ion-beam micromachining); assembly processes; welding	Rolling, finishing mills, specialty steel castings and forgings; engines, turbines; metalworking machinery and equipment; pumps and pumping equipment; air and gas compressors, pressure vessels; tanks, heat exchangers, steam generators, pipes, welding apparatus	Aerospace Metalmachining Mechanical Naval

TABLE 5 *(continued)*

Processes/skills	Materials/products	Potential industrial applications
ELECTRICAL, ELECTROMECHANICAL, ELECTRONICS		
Computing techniques (data processing)	Transducers	Electrical
	Power transmission	Electronic com-
Control theory	equipment	ponents and
Instrumentation		equipment
SYSTEMS ENGINEERING		
Systems analysis		Systems
CAD, CAE		architecture

NOTE: The revisions by Bernardo Jaduszliwer and Simon Solingen in the design of this table are acknowledged. Information has also been extracted from Sábato (1973) and Elkoubi and Bernard (1982). An earlier version appeared in Solingen (1990b).

sis, we are confined to gauge changes in more qualitative terms. Table 5 provides a noninclusive but suggestive description of the processes, materials, and skills affected by the inception of a nuclear industry and of their potential applications in other industries. Such influences and applications include vertical transfers of skills and processes (within state agencies) as well as horizontal transfers to private firms. They reflect changes on products, processes, manufacturing techniques, marketing, and the transfer of labor and managerial talent.[13]

Nuclear Programs and Spillover Effects

Technological change has been associated with expansion in the kind and quality of materials or their range of substitutability.[14] The effects of nuclear programs on materials engineering, metallurgy, and ore processing are therefore a logical place to start. On the one hand, metallurgical skills have been critical to the establishment of a nuclear industry.[15] On the other hand, development of new and superior metals can improve the performance of other important industrial sectors, including machine tools, electric power generation, jet engines, and railroads.[16] Even critics of the nuclear program in Brazil acknowledged a few gains in this area, specifically in the production of high quality steel components.[17] This facilitated, for instance, domestic production of the steel containment shell for Brazil's Angra 3, the second power plant planned under the 1975 agreements but eventually canceled.[18] In addition, while KWU used about 35 different types of steel in its own plants, Brazil narrowed the selection to five in an effort to nationalize the production of tanks and vessels for nuclear plants.[19]

Fuel-cycle activities in both countries induced momentum in the development of other materials. First, uranium prospecting and exploitation compelled the development of geophysics and geochemistry.[20] Second, uranium processing induced new activities in metal mining and processing.[21] Brazil's Poços de Caldas became a successful training ground for geologists and mining engineers.[22] NUCLEMON developed new products to satisfy strong Brazilian demand for heavy minerals that were often procured abroad.[23] It succeeded in filling about 50 percent of national demand for zirconite, whose derivatives are used in optics, glass, chemical and metallurgic industries, and porcelain enamel. NUCLEMON's activities also benefited the sugar and paper and pulp industries. In addition, the uranium concentrate plant could now supply molybdenum concentrate (in the form of calcium molybdate), used in the production of special steels, and, at a later stage, zirconium oxide.[24] Activities in mining, ore processing, and metallurgy resulted in the incorporation of new physical and chemical processes. These included analysis of physical and mechanical properties, metal and ceramic production processes, finishing, and nondestructive testing.[25] Brazil thus began modest exports of technical assistance and training in the area of minerals to other developing countries.[26] Nuclebrás's Center for Development of Nuclear Technology (CDTN) provided services to private firms in mining, steel, metallurgy, chemistry, petrochemicals, hydroelectric energy, hydrology, and environmental sciences.[27]

Technological diffusion from the nuclear industry to other sectors was a priority in the Argentine program. Prominent in ensuring the optimization of this process was CNEA's agency SATI, which provided technical assistance to industry. During its first twelve years (up to 1972) SATI developed a new alloy for welding electrodes, a new process to manufacture tungsten-silver electrodes, a method of analysis of cracks in pressure vessels (also used in the petrochemical sector), and a series of other new products and processes.[28] As early as 1962, six years before approving an order for the first power reactor, SATI signed an agreement with the Association of Metallurgic Industries and has since provided assistance in quality control, development of new products, and expert consulting (a total of 1,092 jobs up to 1983). Over 30 percent of these services were rendered to other state firms. SATI's role was defined with a view toward industrial technology diffusion effects beyond the nuclear sector, and over 50 percent of SATI's trainees were not involved in CNEA activities.

CNEA's dedicated efforts in metallurgy began in 1955 when it created a pioneering metallurgy division, ahead of any other academic or governmental research program in this area.[29] This laboratory did not respond merely to CNEA's own metallurgical needs (particularly in the nuclear fuel area), but also to private sector requirements and to academic research activities at universities and laboratories.[30] By 1971, the metallurgy department had 154 academic experts, 50 percent of them in physics and chemistry.[31] CNEA's Bariloche Atomic Center developed mathematical models on processes of extractive metallurgy. Overall R&D activities in this area resulted in a number of patents and the development of new materials.[32] The metallurgy division handled six hundred consultations from the electromechanical sector in its laboratories and testing services.

I have examined some of the gains (in both countries) in mechanical, electrical, and electronics engineering in the preceding chapter.[33] In light of the technological convergence of capital goods industries, supplying the nuclear sector was expected to diffuse newly acquired technological capabilities to a wide range of productive activities.[34] Brazil's NUCLEP, the heavy components factory, introduced new equipment and machinery (over two hundred machine tools, 50 percent of which were of national origin), built a roll-on/roll-off maritime terminal (at Itaguai) with a 1,000-ton capacity, and developed personnel skilled in welding, pipe fitting, and boilermaking. The acquisition of a large capacity in production of heavy components could then be directed to the production of medium- and large-scale turbogenerators, but, as we have seen, private producers opposed NUCLEP's intrusion into this domain.[35] There were some gains in the area of instrumentation and control, despite German predominance in the supply of electronic and control systems.[36] The original estimate of domestic provision of electronic equipment for the third and fourth plants (5 percent of the total) was revised significantly upward by 1984, but these plants were later canceled. In this area, however, the nuclear program itself was affected by developments in the protected electronics sector.

Other potential spin-offs of nuclear industrial activities cut across industrial sectors, as with improvements in managerial skills and procedures, standardization and norms, and quality assurance and quality control. With respect to managerial skills, NUCLEN's responsibility over plant layout, for instance, implied control and coordination among all inputs and services (mechanical, civil, and electrical plant engineering). FURNAS (the electrical utility) required the development of a similarly high level of technical-administrative capacity. Regarding standardization, it is said that: "To produce technology is to produce standards of operation."[37] Standardization and norms aim at rationalization of all steps of production, marketing, and consumption.[38] The multifunctional and therefore interchangeable nature of component parts multiplies the advantages of standardization. Both Nuclebrás's Committee on Technical Norms and CNEN developed technical norms in the nuclear sector.[39] In general, there was no significant effort to screen norms transferred to Brazil for their applicability to local conditions.

The nuclear program provided momentum to the conscious incorporation of quality assurance and quality control techniques, where Brazil lacked a strong tradition.[40] Quality assurance standards play an important role in technological diffusion through "checks and balances" among manufacturers, material and parts suppliers, and engineering and construction services. Quality assurance for the nuclear program was initially the responsibility of the German agency RW-TÜV, but in 1978 the Ministry of Industry and Commerce created Brazil's Nuclear Quality Institute (IBQN)—an independent agency—to assume those functions.[41] IBQN played an important role in advancing industrial standardization and technological controls throughout Brazilian industry.[42] For the nuclear components supplied by the major Brazilian firms there were at least 40 to 50 subsuppliers that required IBQN certification.[43] IBQN's efforts at nationalizing standards (searching for equivalent safety and quality procedures based on local resources, and the like) were limited.[44] Yet Brazil had come a long way from the Bechtel feasibility study of 1973 that had found its firms deficient in testing, thermal treatment, heavy equipment, surface treatments, design capacity, experience with special materials, and practices of quality control and quality assurance.[45]

The major beneficiaries of such interindustry effects are often buyers of similar or related goods in other industries.[46] With very few exceptions, most components supplied by local firms for the nuclear program could be produced for other sectors, particularly power generation and transmission, petrochemicals and chemicals, mining and metallurgy, iron and steel, aerospace and telecommunications. Thus, exposure to the nuclear area—which requires Level 1 quality assurance standards—benefited activities linked to Petrobrás, Embraer, Eletrobrás, and Siderbrás.[47] Since these four together procure 85 percent of domestic materials, parts, and components, many producers and subsuppliers had to adjust to their specifications. Some private firms requested permission from NUCLEN to use the quality assurance manual for production in nonnuclear areas.[48] In general, most firms that I interviewed agreed that eventually the nuclear program helped them satisfy newly imposed requirements from clients like Petrobrás.[49] Yet at least some producers expressed the view that the few gains obtained from production for the nuclear sector had very marginal application in other areas. Most industries require less stringent standards than those used in the nuclear sector. Moreover, rigorous quality assurance standards were already prevalent in many of these industries.[50] Finally, gains in quality assurance were fairly concentrated among a few large- and medium-sized firms involved in the program, leaving most of the sector practically untouched.

Argentina's CNEA invested heavily in power plant design and construction and developed new products and processes, including the detailed design of a self-energized neutron detector (a part of the core control instrumentation) for the Embalse plant. CNEA also developed ten technical norms, later adapted by the Argentine Institute of Rationalization of Materials. It created the National Institute on Non-destructive Testing (INEND) in 1972, in cooperation with the United Nations Development Program, to diffuse quality control standards in industry. Through its procurement practices and INEND, CNEA advanced the independent qualification and certification of operators.[51] This program—the first in Argentina—was later enforced by the most demanding industrial sectors, such as aeronautics.[52] SATI placed a strong emphasis on creating links to other state firms and agencies. By 1973 it cooperated with the state oil enterprise (YPF), electrical utilities, the National Institute of Industrial Technology, the Ministry of Health, the National Council of Scientific and Technical Research, the State Waterworks Service, and the Institute of Mining and Geology. CNEA officials were also involved in evaluation of patent requests with the Division of Industrial Property at the Secretariat of Industrial Development.

The diffusion of technical and managerial skills was also carried out through the physical movement of high-level CNEA personnel into other areas of industry. These professionals moved to firms like ALUAR (Argentine Aluminum Enterprise), ACINDAR (Argentine Steel Industry), SEGBA (Buenos Aires Electrical Utility), TECHINT, and others.[53] Their performance in procurement of equipment and development of local suppliers suggests a strong influence from technical experts trained in CNEA.[54] In the early 1970s engineering and capital goods firms, together with other buyers of medium- and large-sized plants, incorporated quality control techniques advanced by CNEA. The demonstration effect of the nuclear program was particularly evident in other energy sectors, but its widespread influence is apparent in its inspiration of Ley Compre Nacional (1970), which institutionalized the disaggregation of technological packages.[55] This diffusion of new production standards should not be overstated; few industrial sectors require the level of quality assurance and redundancy of the nuclear sector. The costs of developing such capability cannot, therefore, be deducted by considering broader industrial effects.

Training

In the introduction of a nuclear industry, training may be regarded as a mixed side effect, namely one that ought to be considered an essential input as well as an indirect benefit.[56] In evaluating outcomes in this area, quantitative and qualitative aspects of training need consideration. The absorption of nuclear technology is contingent on the availability of a large pool of qualified multidisciplinary human resources. We know from earlier chapters (5 and 6, in particular) that Brazil fell short, at least initially, of recruiting and training the estimated personnel required for that effort.[57] By 1983 the total number of trainees (in managerial, financing, and technical areas) was about 4,800.[58] Yet, seven years after the program's inception (by 1982), only 58 people were trained in R&D activities (roughly 3 percent of the total trained), as opposed to the original projection of about 40 percent.[59] Basic research is assumed to have greater prospects for generating external benefits (including diffusion) then applied and developmental (nuclear) research.[60] Despite calls by the scientific community for increasing the number of M.Sc.'s and Ph.D.'s in physics from 46 (in 1975) to at least 350, training in physics had low priority.[61] NUCLEN trained about 350 engineers (156 on-the-job, at KWU facilities in Germany) between 1976 and 1984, mostly in mechanical components, electrical engineering, instrumentation and control, plant design, and quality control and quality assurance.[62] NUCLEP trained a total of about 50 engineers and technicians between 1978 and 1984.[63]

The total number trained abroad by 1983 was about 500.[64] They accounted for over 90 percent of the total costs of training, but for only 17 percent of the total number of trainees.[65] The difference between potential training costs of a Ph.D. abroad (through National Research Council fellowships), and those of a technician trained by KWU, is striking. Academic training, however, is not equivalent to on-the-job training, which provides a more practical expertise. Critics argued that the program trained mostly technical managers rather than technological experts, and that KWU exercised excessive influence in the selection of trainees, bypassing those with prior expertise.[66] Nuclebrás officials countered those claims by pointing to services rendered by its own trainees in Germany as a measure of their successful training.[67] Eventually they responded to critics by recruiting more experienced engineers and by supervising foreign training more effectively.

The externalities of training stem from the person-embodied character of knowledge.[68] Thus, both the fact that only about 25 percent of the professional force required predominantly nuclear training, and the interdisciplinary nature of nuclear engineering, had the potential of benefiting other sectors through the movement of trainees to ancillary activities. In other words, although Nuclebrás would not have been able to appropriate all the benefits from investment in training, learning would have provided external economies in the future, both for other state firms and private ones. It is not easy to provide systematic evidence about the movement of human carriers of technology across industrial sectors. An approximation of the flight of professionals from Nuclebrás and subsidiaries—mostly owing to low remuneration—can be obtained by looking at the figures for the years 1983–84. About 223 academic and 255 medium-level technicians left the firm between those years, or 478 in all categories.[69] About 50 percent of trainees in instrumentation and control left for other (particularly private) sectors, as well as some among the 191 trained in quality assurance and quality control (which represented about 9 percent of the total number of trainees).[70] In considering these apparent gains, particularly for the private sector, one should not overlook the high social costs of training through the nuclear program.[71] In addition to technological diffusion through the flight of trainees, Nuclebrás sponsored four-week courses from 1974 to 1978, for all industries.[72]

Brazil's failure to integrate the industrial program with academic training programs in related disciplines was perhaps the most striking contrast with the Argentine training program. CNEA coordinated training in Buenos Aires or Bariloche, in nuclear sciences and engineering, chemistry, biology, metallurgy, geology, mathematics, electronics, mining, and mechanical engineering. Two basic premises guided training in Argentina: the need to provide trainees with a broad background and the value of CNEA-trained personnel to other industrial sectors.[73] As can be recalled from Chapter 5, CNEA filled a vacuum in Argentina in the fields of nuclear engineering, physics, metallurgy, uses of isotopes, and other areas.[74] CNEA's technical staff approached 50 percent of its total personnel, of which 17 percent (or about 54 percent of all technical personnel) held Ph.D.'s.[75] Close to 180 doctoral theses were submitted to CNEA in various scientific disciplines. CNEA's scientists and engineers published over 900 articles in international publications, 2,400 internal documents, and 200 documents as training material, whereas Brazil's agencies had no comparable record.[76] The metallurgy department by itself accounted for over 300 publications in scientific journals and for 30 doctoral theses.

CNEA's one-the-job training programs emphasized close supervision to ensure effective learning and defined a required curriculum and objectives. It screened trainees thoroughly for prior professional experience, language proficiency, and employment stability. By 1972 it had trained about 600 people abroad, while hosting 190 foreign trainers in Argentina.[77] By 1984 CNEA had 2,500 foreign-trained professionals in its midst. The exorbitant amount Brazil paid annually per trainee abroad—relative to the economy of a developing country—was not above that of Argentina ($50,000).[78] Efficient training did not offset the high social costs of training for a nuclear industry although, as in Brazil, many professionals and technicians transferred their new skills to the private sector and to educational and research institutions. In fact only about 30 percent of those trained overseas remained at CNEA.[79]

CNEA inaugurated a training center in 1973, a year before the first plant began operation, graduating 100 reactor operators in the first five years. It offered two-week recycling-for-other-industries courses (about 259 in all), training close to 6,000 participants. Other courses recycled 798 researchers and professors.[80] Nearly 500 university graduates had received training in metallurgy by 1973, of which 130 stayed at CNEA, 280 went to the private sector, and the rest to universities. This diffusion of human resources to other sectors was expected to promote new and improved industrial activities, to which I turn now.

Forward and Backward Linkages and Nuclear Exports

In this section I focus on the nuclear programs' impact on stimulating the development of other industries or entrepreneurial activity in general. Theories of "social overhead investment" examine how the introduction of certain industries induce developmental sequences in associated sectors.[81] In Hirschman's terms, investment in social overhead capital is advocated "not because of its direct effect on final output, but because it permits and, in fact invites directly productive activities to come in" (1958:84). The effects or linkages of a given product line are "investment-generating forces that are set in motion, through input-output relations, when productive facilities that supply inputs to that line or utilize its outputs are inadequate or non-existent. Backward linkages lead to new investment in input-supplying facilities (materials, capital goods) and forward linkages to investment in input-using facilities" (Hirschman 1981:65).

In principle it might be argued that the potential of nuclear industries for creating linkages would be greater in developing countries, where many of the affected industries are relatively weak. Figure 4 suggests potential forward and backward linkages of a nuclear industry.[82]

I discussed in Chapter 3 the emphasis of successive regimes in Brazil and Argentina on the development of capital goods. *Backward linkages* were regarded as an important mechanism in this process.[83] The generic difficulties of measuring backward linkages are exacerbated in analyzing resource demands created by the nuclear industry, for which limited data are publicly available.[84] On the basis of Brazil's weak efforts at nationalizing industrial activities in the sector, one might deduce fewer backward linkages there than in Argentina. Still, increased demand for nuclear fuel, steel, iron, capital goods, and engineering services set in motion new industrial activities in Brazil,

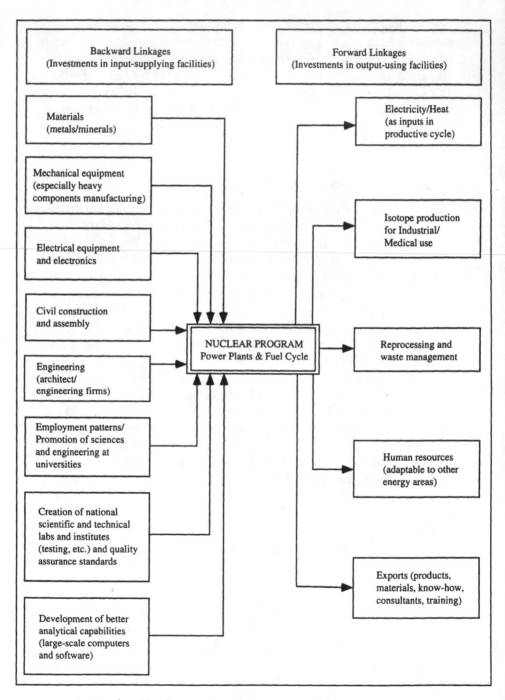

FIG. 4. Forward and backward linkages—nuclear industry.

many of which I discussed earlier in the context of spin-offs. The creation of a new heavy components facility (NUCLEP), with a capacity for handling very heavy equipment unparalleled on the subcontinent, represented a certain expansion of heavy metallurgy capabilities.[85] More significant were NUCLEMON's activities, as well as those of other private and public firms involved in mineral mining and processing, that resulted in a 2,630 percent increase in uranium reserves (between 1975 and 1984), turning Brazil into the fifth largest reserve worldwide.[86] The nuclear industry also required the creation and promotion of services in science and engineering, such as new industrial agents in normalization and quality assurance. The expansion of Brazil's private sector in the 1970s, however, can hardly be associated with the nuclear program.

In Argentina, the industrial diffusion of advances in metallurgy and mineral processing gained in the nuclear sector has been slow and unimpressive. Despite efforts to accelerate backward linkages through forceful procurement of domestic inputs, most components requiring high levels of technological sophistication were still imported.[87] Beyond some fruitful contributions to the expansion of modern manufacturing standards, the nuclear industry was no engine of industrialization, particularly in comparison with the industrial transformations brought about by the automobile sector, for instance. Moreover, the state bore most of the burden of training, at an unusually high price. One-fourth of the $1 million the U.N. Development Program invested in CNEA's metallurgy division could have supported two serious research centers on materials and irradiation.[88] Direct applications of nuclear technology to radiotherapy, materials sterilization by irradiation, and pharmaceuticals could have been obtained from an investment of $100 million to $200 million rather than as the indirect product of a multibillion dollar nuclear power program.

Forward linkages as induced responses to the development of a nuclear industry include electricity and heat generation, themselves inputs in other productive cycles. On the one hand, nuclear-generated electricity had a high degree of substitutability (for alternative energy sources) in Brazil.[89] On the other hand, anticipatory perceptions of abundant energy may have acted to induce new investments. Moreover, the possibility of developing a big nuclear industrial park (including exports) affected investments by Brazil's capital goods and engineering firms, at least initially.[90] The prospects that domestic supplies for the nuclear industry would lead to upgraded technological capabilities and consequently to international competitiveness were considered at the outset.[91] A country's level of technological development is often reflected in the composition of its exports.[92] The regime implanted in 1964 attempted to change Brazil's exports structure by "focusing on increasingly narrower slices of industrial products requiring relatively more sophisticated technology."[93] Nuclear exports should be considered in the context of this shift from exports of raw materials, to manufactures, to sophisticated equipment, to patents/licenses, to person-embodied skills.

Brazil signed scientific and technical nuclear cooperation agreements with many countries, including Venezuela, the People's Republic of China, Argentina, Bolivia, Chile, Colombia, Iraq, the United States, Italy, Israel, Paraguay, Portugal, the Federal Republic of Germany, and Ecuador. Like many other bilateral agreements on scientific cooperation, these are no more than a show of goodwill, with little operational implications. Of greater interest are commercial agreements signed between Brazil and foreign recipients in the areas of reactor technology and equipment and in the nuclear fuel cycle. NUCLEP exported nuclear equipment—in conjunction with KWU—including the ($2 million) lower portion of the reactor vessel for Argentina's third nuclear plant. NUCLEP was also included as subcontractor in tenders for Turkey and Egypt and participated in a direct bid for two nuclear plants in Mexico.[94] These transactions were contingent on KWU's own commercial considerations and leverage and reflected mostly arrangements among KWU affiliates.[95] Overall, private firms did not engage in any meaningful exports of nuclear components.[96]

The expansion of known uranium reserves allowed Brazil to export uranium concentrate (yellowcake) to France and Argentina in 1984 (180 tons and 88 tons, respectively) and to the United Kingdom in 1985 (60 tons).[97] These were partial payments for previous yellowcake imports that did not result in foreign exchange. French and German financing of a new yellowcake plant was repaid in ore currency. NUCLEMON exported over $1 million worth of goods to the U.S., Austria, Netherlands, France, and Japan in 1983, an increase of 222 percent over the previous year.[98] Nuclear exports often packaged different services and equipment. The engineering firm that performed the detailed design of the Poços yellowcake facility—a firm with no prior export experience—won a $280 million contract for the construction of a turnkey yellowcake plant in Somalia.[99] An agreement with Iraq involved the supply of low-enriched uranium, safety technology, and training of Iraqi personnel, in exchange for Iraqi oil.[100] This relationship expanded during the 1980s, with Brazilian provision of technical services in other nuclear-related projects, many of which were of a secret nature presumably owing to their military potential. The Brazilian Mineral Resources Prospecting Company (not a Nuclebrás subsidiary) provided technical assistance to Libya in uranium prospecting, personnel training, and aerial geophysics.[101] Nuclebrás provided technical assistance to Spain in training reactor operators and to various developing countries. Despite this list, the scope of nuclear-related exports in engineering and construction services is quite small, particularly when considering the total volume of exports in engineering and construction services by over 30 firms (about $4 billion between 1977 and 1980 only).[102]

In addition to these nuclear industry–related sales, Brazilian firms involved in the nuclear program exported nonnuclear items. This is where the issue of exports relates to second-order or diffusion effects resulting from exposure to nuclear production. One firm interviewed reported the use of new production techniques in its nonnuclear exports, which included oil pipes to the United States ($60 million) and Belgium in 1984–85, steel pipes for oil and gas lines, and other oil refinery equipment to Australia, Argentina, Mexico, Egypt, Colombia, and certain African countries. Some exports were the result of in-house developments. Another firm interviewed increased its exports of boilers, heat exchangers, and pressure vessels to the United States.[103] Two other firms had very marginal exports. Two engineering firms in the nuclear consortium won bids for detailed design and construction supervision of sections of the Baghdad metro. Others provided services in the area of hydroelectric plants, irrigation systems, telecommunications, and transportation to Chile, Uruguay, Venezuela, Nigeria, Algeria, Angola, and Peru. Export capabilities in construction services existed prior to the introduction of the nuclear program.[104]

This brief overview reveals limited export linkages resulting from the nuclear program. Ironically, those sectors least associated with technology transfer from KWU (such as those engaged in mineral processing and those which supplied Angra 1) appear to have been relatively more successful with exports.[105] No doubt success is highly contingent on international demand for the specific good or service.

Nevertheless, the record clearly suggests that the expansion of Brazil's capital goods and engineering exports in the 1970s was unrelated to the nuclear program.[106]

Argentina had emphasized nuclear exports since the program's inception, and its independent nuclear diplomacy (backed by CNEA's lateral autonomy) stressed a willingness to provide effective training in all aspects of nuclear technology. Argentina signed bilateral nuclear technology agreements in the region with Chile, Paraguay, Uruguay, Bolivia, and Colombia, and with many other countries beyond Latin America. The most comprehensive export project was the construction of the Peruvian Nuclear Institute, including a research reactor and auxiliary installations and laboratories, at a cost of $60 million. CNEA and Argentine producers provided about $30 million in capital goods and technical assistance.[107] In 1985 CNEA agreed to supply Algeria with a research reactor of Argentine design and with enriched fuel, training, and technical support.[108] Another contract for a reactor was signed with Egypt. The private sector and CNEA collaborated in the design of a medium-sized (300–600 MW) nuclear power plant for export.[109]

Expertise gathered in the area of metallurgy allowed CNEA to become, seven years after the inception of its own program, a Latin American training center. The United Nations Development Program, the Organization of American States, and the International Atomic Energy Agency provided support for this center, which trained about six hundred Latin American technicians between 1955 and 1980.[110] A most publicized export item was zircalloy tubing of local design, for the production of fuel elements. There are very few international suppliers of this item, and its transfer is affected by restrictions related to nonproliferation concerns. As early as 1974 Argentina signed an agreement with Col. Muammar Gaddafi's Libya for the supply of prospecting equipment and technical assistance, despite international concerns with this program's military objectives.

Argentine nucleocrats and technicians had served as consultants for the Iranian nuclear program since the 1950s and renewed their nuclear ties with Iran under Islamic leadership. CNEA also provided technical assistance to the People's Republic of China in the negotiation of its nuclear technology transfer agreements, applying its bargaining experience with technology suppliers in exacting favorable terms.[111] CNEA's total exports between 1977 and 1993 did not exceed $300 million.[112] CNEA's legal system of technology transfer is considered a model by other Latin American countries. Argentina's aggressive nuclear export designs often led it to collide with the non-proliferation regime, up to the early 1990s. During the Menem administration the country's foreign policy became aligned with the nonproliferation regime as never before, and the loss of CNEA's autonomy helped bring nuclear exports in line with this new posture and away from an export trajectory hitherto responsive to CNEA's bureaucratic and clientelistic interests.

No discussion of nuclear exports by Brazil and Argentina can overlook the series of mutual nuclear agreements these two countries have signed since the 1980s. In 1980, Brazil's NUCLEP agreed to supply a component for Argentina's Atucha 2 reactor; Argentina leased 240 tons of uranium concentrate (to be returned in kind) to Nuclebrás and supplied $3 million worth of zircalloy tubes to Brazil. In 1985, both countries agreed to cooperate in breeder reactor development and nuclear submarines, among other things. Much of this cooperation was explained as an effort to build a joint nuclear export capability.[113] This policy was formulated under the military regimes ruling both countries until the mid-1980s and was premised on the attractiveness of Brazil and Argentina as nuclear suppliers outside the Western regime of restrictions and control. The international market in nuclear technology (low demand) in the last decade may have done as much to stem export efforts as have political pressures on Brazil and Argentina to abide by rules regarding technology transfers.[114]

The Balance on Spin-offs

This chapter examined some of the industrial-technological spin-offs and linkage effects of nuclear programs in Brazil and Argentina. The programs can be credited with some technological and entrepreneurial gains, including improvement in manufacturing processes, some diversification of allied industries, an increase in skilled personnel, and an upgrading of nuclear engineering skills. CNEA's dedicated efforts in all these areas—stemming from the political conditions examined in Part I—gave Argentina an edge over Brazil by carefully taking training, diffusion, and the creation of industrial linkages into account as more than mere inputs.[115] However, Brazil's incoherent implementation of, and lower sensitivity to, the technological and entrepreneurial dimensions of a nuclear industry did not preclude it from catching up eventually.

In evaluating training as a form of spin-offs, our survey raises a number of suggestions for future studies of the effectiveness of technical training in an industrializing context. First, it is important to identify the political-economic makeup of the sector in order to understand differences in training patterns.[116] Second, it is critical to outline how different training strategies contribute to different developmental goals. Third, the design of industrial programs requires a clear conception of the division of labor between state and private industry in training activities. Fourth, early mapping of the sector's contribution to the formation and diffusion of important industrial skills may steer the training process toward more socially efficient patterns. The most widespread diffusion throughout industry of relatively lower levels of skills may at times be preferred over the sectoral concentration of overskilled labor.[117]

The chapter finds the linkages between nuclear industries and overall industrial development—strongly emphasized by nuclear program advocates in the developing world—to be questionable at best. "Induced developmental sequences," using Fishlow's (1971) term, did not materialize. Most gains were largely confined to industrial "islands of technology," reinforcing the infamous structural duality for which Brazil's economy is known. Brazil has been described as Belindia, a combination of Belgium and India, encompassing a modern sector isolated from a large underdeveloped one.[118] In the aftermath of democratization and liberalizing economic trends in Argentina, there was a recognition even there that the nuclear program contributed little to the country's industrial modernization.[119] CNEA had become a hybrid, stale autonomous agency, a technological enclave in the midst of an untouched environment.[120] The major beneficiaries in both cases were a few large industrial concerns, the upper crust in their respective areas, not necessarily representative of industry.[121] Nuclear activities thus largely reinforced the high industrial concentration of the sectors they touched. In neither case did the nuclear industry create thick linkages to the rest of the economy, as in the case of automobiles.[122] In fact, the nuclear industry was often the recipient of advances in other sectors (notably electronics in Brazil), while siphoning resources from other technological priorities.[123]

In impoverishing other sectors with greater social overhead value, nuclear investments resulted in precisely the opposite of spin-offs.[124] Although, as Fishlow (1971) argues, "the alternative path of growth is largely unknown," opportunity costs cannot be overlooked. These included the relative inattention, for many years, to alternative energy technologies, including hydroelectric and conventional thermoelectric technology, which had greater potential for inducing domestic entrepreneurial activities.[125] Export sectors with potentially greater comparative advantage could have been expanded. Some areas presumed to be advanced through spin-offs could have been advanced by direct technological investment, and at much lower costs. For instance, dedicated efforts in raising quality assurance standards throughout industry could have been undertaken at considerably lower costs than those the nuclear industry required. Thus, the opportunity costs involved in heavy investment in nuclear industries were both short-term (other programs could have been undertaken) and long-term (structural distortions resulted).[126]

Much of this record adds up to something comparable to the experience of industrialized countries that invested heavily in military-related industries. For many years, the prospect of technological spin-offs and entrepreneurial linkages helped maintain political support for gargantuan defense budgets, particularly in the United States. With the decline of U.S. competitiveness, among other things, came a growing recognition that the military perhaps actually aided the erosion of U.S. industrial prowess. It is possible that military industries, spin-offs or not, pushed the other superpower (the former Soviet Union) out of business altogether. At the same time, however, rising industrial competitors in Japan and western Europe used state intervention to promote civilian industries and manufacturing skills with a wider range of applications. In the context of the current worldwide debate over the merits of "picking winners and losers," the industrializing world should find it easy to place the nuclear sector accurately within that spectrum.

CHAPTER 8

Industrial Policy, Technology, and the Political Economy

This book advances the argument that levels of macropolitical consensus and bureaucratic autonomy help explain state intervention in industry. The first section of this concluding chapter summarizes why and how these two variables influence industrial and technology policy. Next the chapter explores the general utility of this argument to understanding sectoral variations—beyond the case of nuclear policy—in Brazil and Argentina. A third section extracts lessons for the analysis of technological change in general and among industrializing states in particular. The chapter concludes with an evaluation of what the Brazilian and Argentine experience with a nuclear industry may portend for other industrializing countries.

Macropolitical Consensus and Bureaucratic Autonomy as Sources of Industrial Policy

The political determinants of industrial and technological policy have come under intense scrutiny, particularly in light of growing trade competition among advanced industrial—as well as rapidly industrializing—states. Up until very recently, the analysis of industrial policy in the industrializing world was segregated from that of their industrialized counterparts. This book proposes an analytical framework that helps transcend that gap.[1] The framework traces the political-economic and technological characteristics of industrial programs to levels of macropolitical consensus within a dominant ruling coalition on the one hand, and to internal institutional characteristics of the state, on the other. The framework is summarized in Figure 1 (Chapter 1).

1. Where macropolitical consensus is high and bureaucratic jurisdictions over an industrial program overlap (that is, where sectoral autonomy is low), the sector will generally reflect such consensus. The importance of sectoral agencies in such cases is diminished, while knowledge of the nature and degree of macropolitical consensus is essential to understand the industrial and technological makeup of that sector. For the most part, industrial policy in the former West Germany, Japan, and a number of East Asian industrializing states (notably South Korea, Taiwan, and Singapore) was typical of this pattern, mostly during the takeoff years.[2]

This model also provided the background for the design of Brazil's nuclear industry in 1974–75. The sector's mutual adjustment between state, private, and foreign entrepreneurship, and the technical option adopted, were responsive to the consensual macropolitical objectives of the military-technocratic regime installed in 1964. These objectives—fast economic growth primarily through exports and macroeconomic balance—were pursued by relying mainly on state entrepreneurship and on foreign technology and capital equipment. A core consideration was the continuous supply of energy-intensive industries (mainly infrastructural and intermediate goods), in which the state played a central role.[3] The nuclear program was thus defined in the hub of central political-economic agencies, allowing the intervention of nonnuclear actors in negotiations and reducing the sector's lateral autonomy. In other words, a more or less consensual hierarchy of macropolitical goals and means, and a horizontally segmented decision making, constrained the range of sectoral options and shaped a nuclear industry in tune with the central parameters of industrial policy.[4] An eroding consensus and a newly gained sectoral autonomy during the late 1970s and early 1980s changed the sector's orientation and increased its latitude to tinker with alternative technical paths. Changes in levels of consensus resulted from the heightened tension between state and private industrialists, stemming from what Barzelay (1986:93) defined as a "contradiction between the social bases of authoritarian rule and the political consequences of heavy reliance on state and multinational firms."

2. Where macropolitical consensus is low and a sector's autonomy high, sectoral agencies are freer to shape an industry almost single-

handedly. Low consensus often implies a tacit tolerance for a widened repertoire of macropolitical options, thus providing an opportunity for autonomous agencies to advance their own institutional interests and preferences in designing their own turf. This pattern was evident in some industrial sectors in India, for instance, following the demise of India's strong postindependence macropolitical consensus by the late 1960s.[5]

Argentina's nuclear program was a paradigmatic example of a maverick agency taking advantage of macropolitical chaos (low consensus) to impose its own institutional agenda. The ability of Argentine political and economic forces to challenge state autonomy intermittently precluded the consolidation of a stable industrialization strategy in the postwar era.[6] The shifts between attempts at macroeconomic balance, inward-looking policies, and their reversal were a clear expression of weak macropolitical consensus. The absence of a consistent hierarchy of objectives broadened the range of options in industrial policy and allowed CNEA's institutional interests, sheltered by the agency's lateral autonomy, to prevail in shaping the nuclear industry.[7] Those interests involved a special emphasis on domestic entrepreneurial and technical resources and were perceived to be best served by heavy water/natural uranium cycles. Domestically produced nuclear components were thus *political* products in Argentina, compatible with CNEA's (and the Navy's) attempt to turn nuclear scientific and technological achievements, and the mobilization of national firms, into political assets in its power struggle vis-à-vis the other two services.[8]

It should be noted, in this context, that the bureaucratic autonomy of sectoral agencies isolated from broad societal goals is neither necessarily more efficient nor socially desirable, and may, at times, involve high social costs. The autonomy of central banks is lauded by some and reviled by others. The German Bundesbank, for instance, has arguably placed sociopolitical and strategic burdens on sections of the German public and on most other European states, while reviving fears of German unilateralism in a Europe poised for unity.

3. Where both macropolitical consensus and bureaucratic autonomy are low, the result is often an incoherent policy formulated and implemented in the context of unstable or cyclical central priorities on the one hand, and of clashing bureaucratic struggles on the other. These conditions are often associated with an intensified search by state agencies for societal clients.[9] The difficulty in for-

mulating and implementing a coherent sectoral industrial-technological program increases exponentially with the number of institutions involved.[10] There is no lack of empirical referents for this
pattern, which requires the analyst to pay special attention to the
role of bureaucratic politics. The United States plunged, in the late
1980s, into one of the lowest levels of macropolitical consensus in
the postwar era, evidenced in debates over different paths to revital-
ize U.S. industry. The Commerce Department's Advance Technology
Program (ATP) and the Defense Department's Defense Advanced
Research Projects Administration (DARPA) were conceived in that
macropolitical context, one that rendered bureaucratic politics and
evolving clientelistic patterns important for the understanding of industrial and technology policy (or its absence) in the United States.

4. Under high levels of both consensus and bureaucratic autonomy, sectoral policy is likely to reflect the model underlying that
consensus, provided sectoral and central government agencies converge on their preferences. The Indian, Japanese, and French nuclear
industries were the result of such convergence.[11] Knowledge about
the makeup of dominant coalitions and their preferences is particularly important here. Where an autonomous agency's preferences diverge from the policy set covered by macropolitical consensus, the
policy outcome may be much harder to predict. The relatively high
level of consensus regarding economic liberalization in the People's
Republic of China in the last decade coexisted, at least in some cases,
with powerful independent sectoral agencies. In the arms industry,
these agencies were able to shape their domain according to their
own institutional interests, even where these were at variance with
central state priorities.[12] However, at high levels of macropolitical
consensus, a coalition can generally be more effective in undermining the power of autonomous agencies, particularly when the maverick agency's choices impose high macropolitical costs.[13]

In sum, levels of macropolitical consensus and sectoral autonomy
help anticipate the explanatory relevance of broad industrial models,
sectoral institutions, or bureaucratic politics as critical research arenas. On the one hand, knowledge about the interests and trajectory
of sectoral institutions can explain the content of certain choices but
not why they prevailed. Understanding outcomes requires us to
specify the structural context within which such agencies operate,
setting limits to explanations based on pure "institutional drive."[14]
In other words, rather than merely positing that institutions are im-

portant, the framework suggests the conditions under which they may be more or less relevant.

On the other hand, identifying levels of macropolitical consensus requires the mapping of relevant state and private interests and power, although sectoral decisions cannot be reduced to such political structures and often bear the imprint of institutional processes as well. In other words, understanding the nature of ruling policy networks is necessary to identify levels of macropolitical consensus but is insufficient in describing choices and outcomes. Samuels's (1987) proposition that state intervention will displace private firms when the ruling coalition is narrow and unstable comes close to linking (low) macropolitical consensus to industrial policy but ignores mediating institutional processes.[15] Argentina's nuclear industry thus departs from Samuels's proposition. Nor did a broader and more stable coalition in Brazil opt for a market-conforming strategy; rather, that coalition chartered state entrepreneurship in the nuclear sector, a policy enforced through a segmented decision-making context. Thus, the political strength of private sector forces does not invariably predict the direction of state intervention. The analysis in Part I depicts an economically more mature and politically more significant entrepreneurial class in Brazil—where private firms were largely displaced from the market for nuclear supplies—than in Argentina, where private firms were cajoled into that market.

Neither is industrial policy always determined by international market constraints and opportunities, as classical dependency studies have postulated. With the erosion of macropolitical consensus in Brazil in the late 1970s, pressures by private national entrepreneurs altered the original division of labor among state, private, and foreign firms.[16] This outcome reinforces the view that greater integration with international capital does not, in Becker's (1984) term, "sap" local industrialists politically.[17] The growing ability of Argentine firms to exact concessions from foreign partners cannot be traced simply to international market opportunities; CNEA's coaching played a fundamental role in realizing that potential, and, in many ways, CNEA prevailed even prior to the onset of such international conditions. As Haggard (1986:366) argues, "dissecting the 'balance of power' within the 'triple alliance' (between international, state, and private industry) demands a theory of public policy."

Private interests are not simply an input into policy decisions. Their behavior is also responsive to prior policy outputs. State inducements and clear commitment can lower the deterring effect of market uncertainty (as in Argentina), or reinforce entrepreneurial risk aversion through weak commitments and strategic (political) uncertainty (as in Brazil).[18] The empirical findings reported here reinforce the view that risk-averse investment behavior is not necessarily a distinctive characteristic of economic actors in industrializing contexts. As with the two cases at hand (Argentina and Brazil), political and market incentives—not technological drive—can best explain the entrance of U.S., European, and Japanese firms into the nuclear market.[19]

Entrepreneurial behavior is thus sensitive to levels of macropolitical consensus and bureaucratic segmentation, which can influence a sector's market and political uncertainty. In the case of nuclear industries, state intervention and mercantilistic objectives were extensive in most cases, evident not only in regulatory activities and reactor choice, but also in the emphasis on nationalization of inputs and in the underwriting of R&D in fuel-cycle and reactor designs.[20] This book's findings thus extend the literature on the nuclear sector, largely devoted to the industrialized world, by considering the experience of nuclear industries in industrializing contexts.[21]

Finally, this study of nuclear industries in Argentina and Brazil provides a window onto "shifting involvements" between state and private power and onto the evolutionary (perhaps cyclical?) redefinition of states as central entrepreneurial actors. The 1990s witnessed the emergence of a relatively high level of macropolitical consensus in both countries, regarding the retreat of the state from what are now perceived to be naturally private economic spheres.[22] Argentina's President Menem has coalesced widespread support for privatization and economic liberalization despite the heavy social costs. Brazil's former president Fernando Collor de Mello's commitment to similar policies won him an approval rating of close to 90 in early 1990. There was a brief hiatus, characterized by lower levels of consensus, between Collor's downfall (on corruption charges) and the election of Fernando Henrique Cardoso to the presidency in October 1994. With Cardoso, Brazil is now considered to be at the threshold of replicating more fully—perhaps "improving" on—the experience of Argentina and Chile in downsizing the state. Among the main beneficiaries of privatization in both countries, one can count a number of firms which were central subjects in our discussion of the nuclear industry. Very few of these firms maintain an interest in the nuclear sector, following the sector's slowdown in the 1980s and eventual collapse in the 1990s.

Consensus, Autonomy, and Bargaining Outcomes

Levels of macropolitical consensus and bureaucratic autonomy can influence the process of bargaining with foreign partners and technology suppliers through their impact on the size of domestic win-sets, on the risks of involuntary defection, and on the credibility of commitments and the reduction of uncertainty.[23] Figure 3 in Chapter 4 summarizes the differential impact of levels of consensus and autonomy on bargaining circumstances.

Low consensus and high autonomy (cell II) often narrow the size of the win-set to the institutional preferences of the sectoral agency. A small domestic win-set, in turn, can be a bargaining advantage, increasing the host country negotiators' leverage over the distribution of gains from the international bargain. This is akin to telling the interlocutor, "No other outcome will be acceptable."[24] An agency's monopoly over ratification and implementation also decreases the risk of involuntary defection. As this study suggests, Argentina's CNEA used all these advantages effectively in negotiations with technology suppliers. The arguably nascent power of Japan's Self-Defense Agency may rely on the currently weak consensus in Japanese politics to secure some of its own institutional preferences, in the context of negotiations with foreign partners over different co-production arrangements in defense systems.

Bargaining advantages can dissipate when both consensus and autonomy are low (cell IV). Such conditions increase uncertainty regarding the contours of the win-set and can raise the risk of involuntary defection and weaken the credibility of commitment. Unstable domestic demands from various quarters (within and outside the state) force foreign partners to continuously assess the balance of forces within the host state. These conditions often result in a protracted and unpredictable bargaining process whereby foreign partners require additional assurances (side-payments) to ensure ratification. Bargaining in the context of Mexico's pharmaceutical (steroid hormones) industry largely resembled this case.[25] In another context, ongoing negotiations between international (public and private) economic donors on the one hand, and the Palestine Liberation Organization (PLO) on the other, reveal many of the negotiating parameters that low consensus and low autonomy often help define: high uncertainty about the contours of the win-set, high risks of involuntary defection, and a weakened credibility over the recipient's commitment.[26]

The independent effects of high consensus seem to be mixed. On the one hand, high consensus within A may dissipate fears of involuntary defection in the opponent B, because there is greater certainty that A "can deliver." In this case, high consensus strengthens A's bargaining position. This situation is compatible with Stepan's (1978) argument that a stronger, cohesive state elite—capable of providing a predictable environment for foreign capital—enjoys a better bargaining position. Manufacturing investors in Singapore, for instance, faced little uncertainty in light of the nature and extent of the host country's consensus over industrial policy. On the other hand, precisely because high consensus foreshadows easy ratification, negotiators from A are less able to use domestic pressures as a bargaining chip to obtain growing concessions from B. The perception of a relatively high consensus in Brazil in the early 1970s weakened the ability of Brazil's nuclear negotiators to extract concessions from KWU on the basis of the need to build domestic support. Paradoxically, as Putnam (1988) suggests, the stronger the state is in terms of autonomy from domestic pressures, the weaker it may find itself to be in international bargaining. Brazil agreed to foreign control of joint ventures, to an array of restrictive clauses imposed by the foreign partner KWU, to suppliers' credits, and to a limited role for its own technological resources.

The effects of high consensus may be better gauged in conjunction with levels of bureaucratic autonomy. High bureaucratic autonomy (cell I), for instance, can increase the risks of involuntary defection, make the environment less predictable, and weaken the advantages of consensus, if there is little convergence between the consensus on the one hand, and the agency's preferences on the other. However, even at relatively high levels of consensus, bureaucratic segmentation (or low sectoral autonomy, cell III) may dilute the clarity of the boundaries of the win-set. Uncertainty about those boundaries increases the risks of involuntary defection, which, in turn, weaken bargaining positions. However, this effect is offset by the rising ability to use domestic dissent as a bargaining chip. India's bargaining with foreign suppliers in the computer sector and Mexico's bargaining in the automobile industry suggest some of the difficulties that a segmented institutional decision making presents, despite a fairly consensual development strategy.[27]

In sum, the conceptual framework advanced here sheds light on how domestic structures and institutions set the terms around which bargaining takes place. Knowledge about the parties' respective levels of macropolitical consensus and bureaucratic autonomy can help estimate the identity of the bargaining agents, the relative bargaining advantages, and the expected transaction costs involved in a given bargaining situation. Information about consensus and about the institutional imperatives of a state agency (that is, knowledge about "structural power") can improve our ability to predict actual outcomes from potential "bargaining power."[28] Moreover, this conceptualization of bargaining can be applied across regime types (including the democratic-authoritarian range) and helps transcend the difficulties of defining strong and weak states as a determinant of bargaining outcomes.[29]

Beyond the conceptual approach to bargaining advanced here, the empirical findings suggest that the objectives of technology recipients are not necessarily compromised by international market or political forces. The external context did not force the choice of reactor technology or a given distribution of costs and benefits among state, private, and foreign actors. Two technology recipients with comparable capabilities reacted differently to a fairly similar set of external challenges and opportunities, seeking different levels of indigenization of equipment and technological inputs.[30] Despite these differences, recipients were able to extract concessions from suppliers in both cases. Grieco's (1984, 1986) and Encarnation's (1989) emphasis on host countries' ability to take advantage of competitive pressures among technology suppliers proves accurate even for products as highly sophisticated as nuclear reactors.[31] Even where such opportunities are foregone during negotiations, as was largely the case in Brazil,[32] host countries can erode the supplier's prerogatives through effective technology absorption. Finally, diversification of technology sources can reduce a recipient's commercial and *political* vulnerability; this diversification is thus an important instrument for recipients to counter restrictive clauses and overpricing.

For their part, technology suppliers pursue the control of joint ventures, even as minority partners. That, in turn, enables them to maximize their own share of inputs—often under the cloak of quality assurance and reliability considerations—at the expense of available domestic equivalents. Suppliers also seek maximum levels of return on their technological edge, use suppliers' credits (which undermine domestic sourcing), and are anxious to slow down—through restrictive clauses—the ability of recipients to compete with them.[33] These efforts can impinge on the development of local technological capabilities in the short term. However, technology recipients can respond with unpackaging strategies and effective screening to offset the negative effects of suppliers' terms.

In short, external opportunities are not always taken, whereas constraints can be overcome.

Understanding Intra- and Extra-Sectoral Variation

State intervention is neither consistent over time nor across sectors. State agencies overviewing different industries exhibit different "elasticities" (relative autonomy) vis-à-vis political, economic, and social pressures, and vis-à-vis other state agencies. Levels of macropolitical consensus and bureaucratic autonomy can help uncover sectoral variations of high relevance to our understanding of industrial and technology policy.[34] This points to a logical next step with which to follow this study, a step that would explore cross-industrial differences in light of variations in consensus and autonomy. What follows is no more than a stylized, suggestive application of this book's central argument to other sectors. This preliminary review of selected industries in Brazil and Argentina highlights the utility of the conceptual categories of consensus and autonomy for explaining not only varying mutual adjustments among foreign, state, and private actors but also technological behavior and bargaining performance.

SECTORS EMBODYING BRAZIL'S GENERIC MODEL (AND THE ABSENCE OF ARGENTINE COUNTERPARTS)

Under conditions of relatively high macropolitical consensus and low lateral autonomy, sectoral policy is expected to resemble the generic industrial model—much as Brazil's nuclear decisions of 1974–75—a situation characteristic of cell III in Figure 1. In effect, the primacy of Brazil's 1960s model was evident in many of the industrial areas mapped by central planning (rather than by autonomous sectoral agencies). Secure energy and intermediate goods supplies were key to the model's success, even prior to 1973, and state enterprises were entrusted with this task. The scale, risk, capital, necessary coordination—and, above all, urgency—required in these infrastructural and intermediate activities were perceived to exceed the capacity or incentives of private capital.[35] The expansion of state entrepreneurship was particularly evident in the case of Eletrobrás (the public electrical utility), Petrobrás (the state monopoly over oil exploration and refining), and Companhia Siderúrgica Nacional (steel), among others.[36] In the year 1974 President Geisel approved the construction of two gigantic hydroelectric plants (Itaipú and Tucuruí) and began negotiating the nuclear agreements with KWU. In 1975 his government created Nuclebrás and authorized Petrobrás, in an unprecedented move, to sign risk contracts with foreign firms for oil prospecting.

In fulfilling their main mission, these companies imported equipment liberally and established joint state-foreign ventures, often displacing private Brazilian producers, much in the same pattern as Nuclebrás.[37] Eletrobrás emphasized large-scale projects (Itaipú, Tucuruí, Ilha Solteria) in the 1970s, involving advanced technology and imported machinery.[38] Suppliers' credits and packaged foreign equipment became a generalized feature in the power plant sector, where multinational subsidiaries dominated the domestic market, increasing the massive levels of imports of materials and components.[39] Such credits also accounted for a variety of expensive equipment imported for the steel railway (Ferrovia do Aço), thermal, and steel plants. Liberal import policies contributed to the idle capacity of national turbine and hydrogenerator producers.[40] Although Petrobrás's initial orientation in the 1950s had been to strengthen domestic productive and technological activities, this mandate was relaxed in the 1960s and 1970s, in the direction of improving efficiency in energy supplies. Petrobrás increased its emphasis on exploration; domestic production since 1973 led to demand for imported equipment, particularly for rigs for offshore exploration.[41]

State enterprises in the steel sector, such as Companhia Vale de Rio Doce (CVRD) and Companhia Siderúrgica Nacional (CSN), created engineering subsidiaries in association with multinationals, along the lines of Nuclebrás's NUCLEN. In many cases these ventures and others in the steel sector (such as Siderbrás, created in 1973) competed with national private firms.[42] Private firms resisted the development of competing basic engineering subsidiaries within state enterprises. State agencies also compelled domestic producers to rely on foreign technology and licensing. During the 1970s the volume of spending for foreign technology was two to three times that spent on indigenous R&D, exactly as in the nuclear sector.[43]

This expansion of state production and multinational activities initially built on—but ultimately helped crack—the old macropolitical consensus by the late 1970s. Private Brazilian entrepreneurs were no longer satisfied with the role assigned to them in the "triple alliance." As pointed out in Chapter 3, the nuclear program became a lightning rod in the industrialists' assault on statism, and their response spread to other sectors. Industrial equipment producers protested the import of turbines and alternators for Itaipú (the world's biggest hydroelectrical dam), Tucuruí, and others.[44] The Second National Development Plan thus imposed domestic sourcing of capital goods for infrastructural projects as a target.

Private supplies of equipment for the electrical sector increased with state inducements such as active financial support by the National Development Bank (BNDES) for projects with an equipment nationalization index of over 85 percent. In 1980 Brazilian equipment producers blocked Eletrobrás's attempt to purchase equipment for hydroelectric plants (in exchange for agricultural goods) in order to speed up the completion of two stations.[45] The regime's attempt to define a new consensus (by getting important industrialist constituencies on board again) encountered no feeble resistance among some Siderbrás state enterprises. Yet by 1979 Siderbrás and CVRD introduced programs supported by FINEP that emphasized domestic supplies and national design of steel and mining plants. Petrobrás's efforts to develop national technologies led from Brazil's total dependence on imported technology for offshore oil production in the mid-1970s to exports of engineering services and equipment in this area (by private firms) in the mid-1980s.[46]

In sum, this overview highlights the similarities between these sectors and many of the characteristics of Brazil's nuclear program, emphasizing the hegemony of central priorities—backed by the strong macropolitical consensus of the early 1970s—particularly in infrastructural and intermediate areas. In some cases, sectoral agencies and state-owned companies enjoyed significant levels of lateral autonomy. For the most part, as with Petrobrás, the interests and trajectory of these institutions converged with the overall model of industrialization.[47] Where there was no convergence, central priorities often prevailed over sectoral designs. This seems to have been the case, for instance, with CEME (Central de Medicamentos), which was granted significant lateral autonomy in 1971. Its efforts to develop national technology ran aground, and in 1975 CEME was removed from the Office of the Presidency, ending its bureaucratic autonomy. Its research functions were transferred to the Ministry of Industry and Commerce, and distribution was placed in the hands of the Ministry of Welfare.[48] Finally, this overview also reflects the impact of an eroding consensus in the late 1970s, which created new adjustments among state, private, and foreign capital, and new patterns of procurement and of bargaining for technology.

In concluding this section reviewing sectoral developments under higher levels of macropolitical consensus, the absence of Argentine cases (cell III in Figure 1) is noteworthy. As argued extensively in this book, the postwar history of Argentina is primarily characterized by record-low levels of consensus. The emergence of a liberalizing and privatizing consensus in the early 1990s offers new ground for further testing the arguments offered here, even as the process unfolds.

MAVERICKS IN ARGENTINA AND BRAZIL[49]

Under conditions of low macropolitical consensus and high lateral autonomy—as with Argentina's CNEA—we can expect sectoral agencies to guide a sector's industrial strategy almost single-handedly. An overview of the oil, steel, petrochemicals, and weapons industries in Argentina, and of informatics, aircraft, and arms production in Brazil, confirms such expectations.

The four Argentine sectors were part of the Army's industrial fief-dom, which mirrored the pattern of independent Navy control of CNEA established in the 1950s. The lateral autonomy of each service, in the midst of Argentina's macropolitical chaos, allowed them to leave their institutional imprint on their respective industrial domains. The Navy, we recall from Chapter 2, aimed at maintaining its otherwise eroding power vis-à-vis the two other services by co-opting important clienteles into CNEA and by exuding technical competence and professionalism.[50] The Army had acquired unchallenged status among the three services, particularly since 1963, and had dedicated itself to expanding its control over new industrial sectors.

State entrepreneurship was at the heart of this strategy of horizontal expansion into new industries and inhibited effective state subsidiarity, which was the Navy's guiding principle. Quite often, state firms under Army control displaced the private sector from markets this sector was capable of, and interested in, serving.[51] This was evident in oil (Yacimientos Petrolíferos Fiscales—YPF), steel, petrochemicals, and in the complex of civilian and military firms associated with the Directorio General de Fabricaciones Militares (DGFM—General Directory of Military Production).[52] The Army invested little in substituting imported equivalents for national technology and equipment throughout its industries.[53] This behavior is particularly striking because the Army captured 18 percent of central budget allocations for R&D, in contrast to the Navy's 2 percent.[54] The Army's R&D professionals never acquired the power of their CNEA counterparts because they were much less central to the Army's institutional interests and technological trajectory.[55]

Technical achievement and state subsidiarity had been the political pillars on which the Navy's CNEA relied for institutional survival, within its relatively small industrial niche; neither strategy was central to the Army's drive for expansion over a wide range of industrial sectors. Notably, the very fact that sectors with military applications (conventional, nuclear) differed in their industrial characteristics and technological orientations suggests that military potential per se does very little to explain variations among such sectors, with respect to state entrepreneurship and technological effort.[56] This variance may be counterintuitive and challenges more conventional interpretations of Argentina's nuclear program.

Brazil's informatics, aircraft, and arms industries were consolidated in the late 1970s and early 1980s, under conditions of lower macropolitical consensus and higher sectoral autonomy than those identified for the nuclear sector. To some extent, the sources of such autonomy were, much as in the Argentine case 30 years earlier, a division of industrial niches among the armed forces; this process became more pronounced in Brazil only in the 1970s.[57] Independent sectoral agencies were able to impose, in differing degrees, relatively coherent—albeit not always successful—industrial strategies in these sectors. In all three cases, their strategies evoked the clientelistic (market-conforming and market-creating) orientation of Argentina's CNEA, with its emphasis on domestic production experience and research activities, import controls, and state financing. It is never easy to argue the counterfactual path of history that, in this case, would posit that these sectors might have looked quite different had they been consolidated—as Brazil's nuclear industry—in the late

1960s or early 1970s. It is not at all speculative, however, to point out that under the new structural and institutional conditions of the late 1970s to early 1980s, these sectors bore the trademark of forceful sectoral agencies capable of steering through incoherent macropolitical objectives.

Early efforts in the development of a computer industry resulted in the creation of a state-owned joint venture, COBRA, in 1974, following a pattern of expanding state entrepreneurship rather than private markets. A major source of difficulties for COBRA (and its state-owned predecessor Digibrás) was the absence of a sponsoring ministry with powers to protect and regulate the home market until the late 1970s. Conceived under the conditions of the early 1970s, COBRA was stillborn.[58] By the end of the decade, against the broader background of an anti-*estatização* campaign, COBRA's hope to control the domestic market faded with the entry of private Brazilian competitors. The Navy's growing autonomy allowed the creation, in 1979, of the Special Secretariat for Informatics (SEI) as a protectorate charged with developing Brazilian hardware and software industries and processing capabilities through the purchase of domestic equipment and services.[59] The 1984 Informatics Law excluding foreign firms from the Brazilian market until 1992 became SEI's major policy instrument.[60]

SEI's direct subordination to the National Security Council ensured its independence to control and coordinate all activities in this area, including state procurement, services, and development of a human infrastructure. Protection through "market reserve" attracted powerful allies in the banking sector (a shareholder and important client), sections of the military and the foreign policy community (Itamaraty), scientists and technologists (particularly through its R&D arm at Campinas), and private producers.[61] The outcomes of this policy from a technological perspective were controversial and of less concern to us here than the policy's sources. Beyond electronic systems, in the 1980s the Navy expanded its control over communications and an array of nuclear activities geared to develop Brazil's first nuclear submarine.

The Air Force developed the state-owned firm EMBRAER in 1969 as the national champion of the aircraft industry, out of its Centro Técnico Aerospacial (CTA), the locus of technological research in aircraft design and production. At this point, EMBRAER followed the general model of expanding state activities where private firms were thought not up to the task.[62] The Ministry of Aeronautics manipulated Brazil's domestic market for civilian and military aviation to EMBRAER's advantage, through its procurement power, R&D support, and protective tariffs. The Ministry had not only effective control over EMBRAER itself, but also increasingly concentrated— throughout the 1970s—R&D, training, financial, fiscal, marketing, regulatory, and international bargaining (for technology) functions related to the sector. By the late 1970s, the Ministry managed to camouflage EMBRAER as a mixed enterprise, as a result of a tax incentive scheme that provided the firm with low-cost, long-term, intervention-free capital.[63]

In 1979 EMBRAER was the sixth largest aviation company in the world (outside the United States), producing and exporting a variety of planes (air frames, engines, parts, and navigation equipment). Emphasis on local designs implied a heightened role for Brazilian scientists and engineers, now part of EMBRAER political constituencies. Some models were the product of skillfully negotiated industrial cooperation agreements with foreign suppliers, designed to achieve rapid market penetration without excessive technological dependence.[64] Domestic sourcing helped diffuse technological capabilities to three hundred suppliers, in a rather more successful attempt at backward-linkage creation than was the case for the nuclear industry. Finally, the Ministry of Aeronautics nurtured the private firm Avibrás in missile technology.

IMBEL came into being in 1975 under Army sponsorship as a state-owned enterprise, in a pattern compatible with the reigning consensus at the time. However, the anti-*estatização* movement of the latter part of the 1970s and an extant private infrastructure in automobile manufacturing acted as effective barriers against state expansion in this area, particularly by the end of the decade. By the early 1980s, IMBEL was a semiprivatized cooperative structure among state (particularly Army) and 55 private firms. It exported over 90 percent of its production, including armored vehicles, sophisticated rocket systems, and missiles, to 50 countries.[65] IMBEL relied on national designs (about 17 percent of its sales were invested in R&D, $1 million in 1980) or on carefully selected and negotiated coproduction agreements with several suppliers. The sector's performance reflected coordination of fiscal, industrial, and explicit scientific and technological objectives.[66] A private firm, ENGESA, grew to become the world's largest producer of armored vehicles in the

1980s, exporting to over twenty countries. The export successes of the arms production sector ended with the sharp contraction of international demand on the one hand, and the heightened levels of supply on the other, following the end of the Cold War.

Technology, Learning, and the State

Some of the findings of this book regarding the process of technological change may be relevant only to high technology industries characterized—as the nuclear sector—by a high level of sophistication, intensive R&D, long lead times, political-strategic constraints, and specialized by-products. Such sectors may include space and telecommunications, aircraft, and oil exploration and processing. Other findings, however, bear on a wider range of industrial activities, pointing to the value of studying the political economy of technology "in the small."[67] As the literature on the microeconomics of technological change suggests: "Contemporary attempts to uncover the costs and benefits associated with different paths of technological development are . . . hampered by the paucity of adequate case histories."[68] This particular study of the nuclear sector may thus contribute, in general terms, to the conceptualization of technological change as a function of macropolitical and institutional determinants.

On the one hand, Argentina's experience with the absorption of nuclear technology under conditions of strong bureaucratic autonomy reinforces the view that state inducements in the form of reliable procurement patterns, fiscal incentives, protection, and underwriting of training and equipment costs *do* make a difference in the rate of technological development and absorption. This evidence suggests, as many firm-level studies have now recognized, the need to come to terms with exogenous factors influencing a firm's willingness, effort, and achievements in technological learning.[69] Consistent state intervention in technology generation (through R&D efforts) and in diffusion of technical information deepened the scope of technological activity in Argentina. But within Brazil's heavily segmented bureaucratic framework, and against a background of weak technology transfer regulations, delays in the coordination of financing and fiscal incentives, limited efforts to support the technological development of national private firms, and poor coordination with research centers, technology suppliers were able to restrict technology transfer to the bare minimum, to force unnecessary licensing and restrictive clauses, and to undermine efforts to integrate domestic research institutions with industrial activities.

Beyond the obvious fact that state intervention matters, this study suggests that dissimilar state structures provide disparate sets of incentives and risks to private entrepreneurs. The state's ability to signal its intentions efficiently is highest in cases where the sectoral agency's autonomy is high. This is so even if macropolitical consensus is low, because such agencies are often able to cushion private investors from uncertainty. Argentina's CNEA lowered the deterring effect that strategic (political) and market uncertainty can have on the technological behavior of private entrepreneurs. In other words, Argentine nuclear investors committed their resources to *political* products and were less exposed to the pervasive market uncertainties of postwar Argentina.

Signals may be efficient if both macropolitical consensus and bureaucratic autonomy are high, particularly if a happy convergence reigns between central agencies (broad industrial patterns) and sectoral ones. Under low bureaucratic autonomy and low macropolitical consensus, signaling is least efficient, because the outcome of bureaucratic struggles is often hard to predict. Low autonomy and high consensus may allow more efficient signaling than in the previous case, although a high number of agencies may inhibit such efficiency. As Brazil's nuclear sector suggests, a segmented arena of policy implementation can lead to difficulties in coordinating regulation, procurement, and financial and fiscal incentives, thereby increasing economic and strategic uncertainty for entrepreneurs.

On the other hand, Brazil's experience suggests that technological change can come about as a consequence *pari passu* of the technology transfer process, even without dedicated efforts at absorption.[70] Clearly, a concomitant investment in local R&D can reduce the disadvantages of licensing, particularly regarding technology absorption, obsolete technology, and restrictive practices. Yet the contention that merely (passively) absorbing "production techniques" or "detailed design" skills does not lead to basic design capabilities obscures the fact that an independent technology-*learning* capacity has proved to be more important than (in fact, is a precondition for) an independent technology-*creating* capacity.[71] Learning allows successful subsequent searches for technology, adaptation and improvements of foreign technology, and the ultimate development of a domestic design capability. "Learning by training" demands a considerable educational effort on the part of state and private actors. Private firms can rely on "learning by hiring," reaping the benefits of state investments in training of highly qualified people. Finally, "learning by using" can result in positive feedback to designers and producers.[72] Learning is particularly effective when considered as an essential *input* rather than as a potential *side-effect*.[73]

The putative decline of U.S. competitiveness has called attention to the importance of manufacturing skills in technological development in today's international economy.[74] Laura Tyson, for instance, argued that high technology industries create "a virtuous cycle of economic growth, innovation, and learning by doing" that stands at the heart of global competitiveness.[75] The development of research cooperatives and government-funded research centers in Europe and Japan highlights the growing relevance of states in reinvigorating technological capabilities in industry. The Clinton administration's initial plan for industrial renovation included the creation of manufacturing extension centers designed to diffuse extant technical capabilities throughout industry, the development of incentives to encourage private sector investments in areas like information technologies, and the attempt to use the reservoir of scientific and technological expertise from federal laboratories.[76] Without elaborating on their substantive merit, one can recognize, in some of these programs, elements of the "Sábato triangle" and of the classic technology policy considerations of industrializing states. Thus the new global economy compels the need to merge a vast literature on industrial and technology policy that has hitherto segregated the industrialized world from the industrializing experience.

Nuclear Industries, Technology, and Industrialization

Beyond the conceptual and empirical issues discussed in this chapter so far, what other general lessons can we learn from the experience of Argentina and Brazil in the development of nuclear industries? Taking stock of the balance of costs and benefits has special relevance at a time when other industrializing states might be considering investments in this area.[77]

First, prospective newcomers may be well served by a proper assessment of overall energy endowments and of the place of nuclear energy within them. In this study, I have referred to Argentina's and

Brazil's energy alternatives only tangentially because my main purpose was to analyze the two countries' *choices of nuclear industrial structure and technology* rather than why these countries sought nuclear energy in the first place.[78] It is nevertheless important to stress that neither program followed an obvious or impeccable economic rationale, given the availability of less expensive sources of energy in both cases. It is equally pertinent to recognize that neither was energy the prime consideration in the development of European nuclear industries in the 1950s and 1960s. Despite declaratory statements emphasizing energy needs, either the securing of world market shares in advanced technologies, or the military potential of nuclear activities, or both, were the most prominent factors explaining the gravitation of these countries toward nuclear industries.

Second, despite inappropriate popular characterizations of Brazil's nuclear program as "technologically dependent" and Argentina's as "self-reliant," both countries relied on foreign reactor technology, as was the case for most nuclear programs in the industrialized world.[79] Neither of the two different technical-industrial paths they adopted were *technologically* independent, although, as we have seen, diversifying the sources of technology can reduce a recipient's commercial and *political* vulnerability.[80] At the same time, both cases also reinforce what we know from other contexts (most recently, Iraq): that mastering nuclear fuel-cycle technology is not beyond the reach of any state at medium-level stages of industrialization. Such efforts, however, do not require the development of a large-scale *energy-bound* (rather than security-related) program. The bottom line from a global nonproliferation perspective—one largely absent in this book—is that bilateral and multilateral political efforts to outlaw nuclear weapons seem a more effective solution than efforts to control technical exports which might, at best, delay technical mastery of nuclear fuel-cycle capabilities.[81]

Third, both the tendency to identify scientific and technological development with large-scale, capital-intensive projects, and politicians' expectations of political externalities from such projects, are not unique to developing countries or authoritarian regimes. Yet policies of highly concentrated benefits can be more easily pursued where there is little political accountability. The military-technocratic regimes of Brazil and Argentina in the 1960s and 1970s were particularly prone to megalomaniac designs in the space, weapons, and electronics industries, among others. Nuclear programs became a classic example of "technological fixes" that, rather than extricating these countries from technological and socioeconomic bottlenecks, absorbed a far larger share of resources than the public benefits they were able to provide. It is quite clear that the major beneficiaries of the Brazilian and Argentine nuclear industries were a restricted number of self-contained state agencies and corporations, a few large private firms, and a narrow strata of professionals and highly skilled technicians. A positive impact on redressing the distortions of the labor market can hardly be expected from programs where capital costs represent over 70 percent of the project.[82] The programs generated not only few domestic jobs but also absorbed a skewed proportion of highly skilled people into an industry that proved to be far less than an industrial "winner."

Fourth, and directly related to the previous point, our probing of direct and indirect industrial *technological* effects of nuclear programs reinforces what some economists suspected (and what both superpowers found out toward the end of the Cold War era): that technological spin-offs do not easily, inevitably, and consistently flow out of "privileged industries" into other markets.[83] Gargantuan investments in military-related sectors ended up weakening, not improving, the international competitiveness of civilian U.S. industries, while siphoning resources from the broader industrial, educational, and communications infrastructure.[84] Some developing countries have similarly associated nuclear industries with the setting in motion of "compulsive sequences" throughout the industrial structure. Argentina, in particular, emphasized the nuclear industry's potential to induce backward linkages through domestic procurement of equipment and services.

The outcome of CNEA's dedicated efforts in this direction suggests that the nuclear sector may hold little promise as a "heartland" technology, particularly when compared to automobiles and microelectronics.[85] Not only were multiplier effects quite limited in both Brazil and Argentina, but they also absorbed resources from sectors with greater developmental potential. The observed linkages stimulated investments in ancillary technologically complex, capital-intensive areas with little interindustrial applicability (mostly in fuel-cycle activities and nuclear-related equipment). The "baroque" requirements of such equipment, for instance in degrees of durability and performance, placed a barrier on their transferability to cost-sensitive industries. Clearly, the smaller the spin-offs and industrial linkages of nuclear programs, the larger the real (as well as the opportunity) costs to the economy.[86] It is possible to strengthen scientific, technological, and entrepreneurial capabilities *directly* at much lower costs than via programs of this nature. The methodological difficulties involved in analyzing (let alone predicting) second-order effects place a warning sign on the tendency to assume the windfalls associated with high technology projects without question.

In sum, at the very least, the evidence from this study of direct budgetary and opportunity costs suggests the need for developing countries to reevaluate their view of nuclear technology as a developmental panacea.[87] Moreover, the relationship between an industrial-technological activity and the sociopolitical nature of a regime flows in both directions. On the one hand, the political-economic context shapes industrial sectors. On the other hand, the outcome of that intervention affects the legitimacy and political longevity of ruling coalitions and of the particular consensus they bring to bear on policy matters. Therefore, the attempt to extricate technical considerations from effective development, equity, and general welfare considerations may not be merely normatively undesirable. The sweeping democratization of the industrializing world has rendered it politically unwise as well.

Appendixes

The Nuclear Fuel Cycle

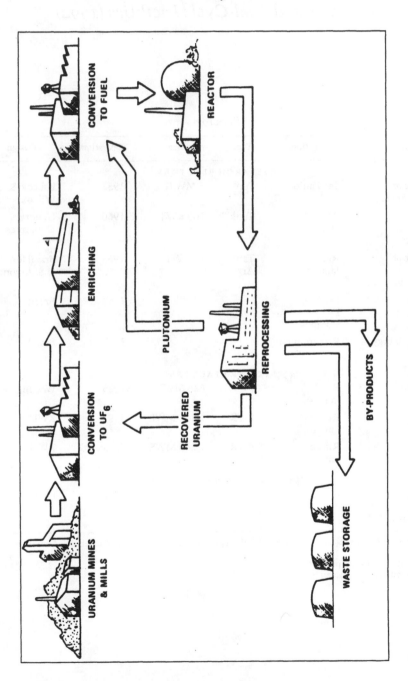

Reprinted from *The Nuclear Industry*, U.S. Atomic Energy Commission, 1978:14.

Brazil's Nuclear Program: Reactors and Fuel-Cycle Facilities (1994)

Reactors

Site	Location	Type	Size	Operation	Design
RESEARCH REACTORS					
Institute of Atomic Energy	São Paulo	Pool	5 MWT	1957	Babcok & Wilcox
Institute of Radioactive Research	Belo Horizonte	Tank	100 KWT	1960	General Dynamics
Institute of Nuclear Engineering	Rio de Janeiro	Tank Arg.	10 KWT	1965	Brazil (on U.S. Argonne)
Aerospace Technology Institute	São Jose Dos Campos			?	Brazil
Nuclear Energy Center	Pernambuco			?	Brazil
POWER REACTORS[a]					
Angra 1	Rio de Janeiro	PWR	626 MWE	1985	Westinghouse
Angra 2	Rio de Janeiro	PWR	1,245 MWE	1997[b]	KWU
Angra 3	Rio de Janeiro	PWR	1,245 MWE	Canceled	KWU
HEAVY COMPONENTS FACTORY					
NUCLEP (joint venture: Nuclebrás-KWU)	Itaguai			1980	KWU

SOURCES: *FBIS*, 1978–95; *Nucleonics Week*, 1978–95; *Relatório*, 1983, Nuclebrás, *Annual Reports* (1983–86).
[a]Six additional reactor projects canceled.
[b]Estimated date of completion.

Fuel-Cycle Facilities

Facility	Location	Basic design	Operation
Uranium mining and concentrate (yellowcake)	Poços de Caldas	Pechiney Ugine–Kuhlman	1982
	Itatiaia pilot plant	Brazil (?)	1986 Not operating
Conversion (UF_6)	Resende	Pechiney Ugine–Kuhlman	Postponed
	IPEN (São Paulo) pilot plant	Brazil	1984
Enrichment			
Jet-nozzle	Resende demonstration plant	NUCLEI (Steag, Interatom, Nuclebrás)	Canceled
	Resende industrial plant		
Ultracentrifuge	IPEN pilot plant	Brazil	Operating
Laser	CTA	Brazil	Under development
Fuel element assembly	Resende	KWU-RBU	1982
Reprocessing	Resende pilot plant	KEWA UHED	Canceled
	IPEN São Paulo	Brazil	1986

Argentina's Nuclear Program: Reactors and Fuel-Cycle Facilities (1994)

Reactors

Site	Location	Type	Size	Operation	Design
RESEARCH REACTORS					
RA 0	Córdoba	CE	10 W	1958	CNEA
RA 1	Constituyentes	MA	150 KWT	1958	CNEA
RA 2	Constituyentes	CE	30 KWT	1966	CNEA
RA 3	Ezeiza	Tank	5 MWT	1967	CNEA
RA 4	Rosario	SH	0.1 W	1971	CNEA
RA 6	Bariloche	Tank	500 KWT	1982	CNEA
RA 8	Pilcaniyeu		70 MWT	?	INVAP
POWER REACTORS[a]					
Atucha 1	Buenos Aires	HWPV	367 MWE	1974	Siemens AG
Embalse Río III	Córdoba	HWPT	648 MWE	1983	AECL
Atucha 2	Buenos Aires	HWPV	745 MWE	1997[b]	KWU

SOURCES: *FBIS*, 1978–95; *Nucleonics Week*, 1978–95; *Interciencia* 4, no. 5 (1979): 301; CNEA, *Annual Reports* (1978–88).

NOTES: "RA 5," not included in this table, is now the power reactor "Atucha 1."

ABBREVIATIONS: CE = critical experiment; MA = modified argonaut; SH = solid homogeneous.

[a]Three additional plants in planning stage.

[b]Estimated date of completion.

Fuel-Cycle Facilities

Facility	Location	Basic design	Operation
Uranium mining/ milling	Malargüe	CNEA	n.a.
	Sierra Pintada	CNEA	1984
	Los Gigantes	CNEA	1982
Purification (UO₂)	Córdoba	KWU-RBU	1982
	Córdoba	CNEA	1983?
Heavy water (D₂O)	Zárate (pilot)	CNEA	1984?
	Arroyito (production)	SULZER	1993
Fuel element assembly	Constituyentes (pilot)	CNEA	1976
	Ezeiza (manufacturing)	KWU-RBU	1982
	Ezeiza (production)	CNEA	1986?

Fuel-Cycle Facilities

Facility	Location	Basic design	Operation
Zirconium and zircalloy	Bariloche (pilot)	CNEA	1979
	Ezeiza (production)	CNEA	1984
Enrichment (gas diffusion)	Pilcaniyeu (pilot)	CNEA	1983
	Pilcaniyeu (production)	CNEA	1995?
Reprocessing	Ezeiza (lab scale)	CNEA	1967
	Ezeiza (pilot)	CNEA	1989 suspended

Nuclear Power Plant (Pressurized Water):
Schematic Diagram

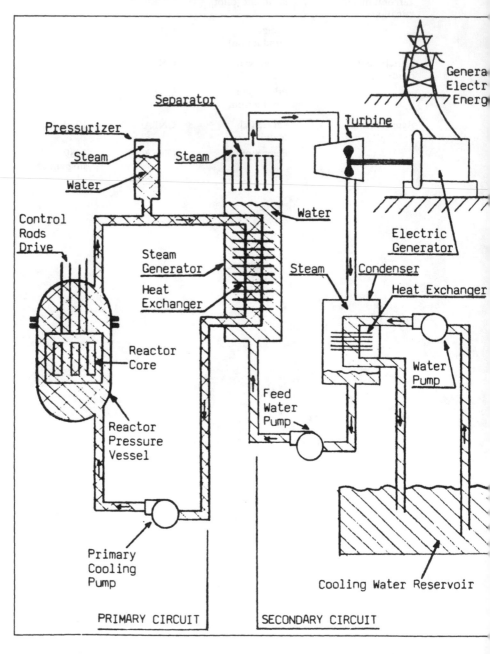

Reference Matter

Notes

Chapter One

1. The share of budgetary spending on state enterprises in Brazil and Argentina at that time—about 11–12 percent—was almost equivalent (B. Schneider 1991).

2. Samuels differentiates among three basic forms of state intervention: market-conforming (where the state reproduces market structures by fortifying the position of existing firms, mainly through financial, regulatory, and R&D incentives), market-displacing (where it becomes an entrepreneur, impinging on private interests), and market-creating. On market-conforming state intervention, see Johnson (1982).

3. On state entrepreneurship in the oil sector, see Klapp (1987).

4. Scheinman (1972); Kitschelt (1982 and 1991); Campbell (1988).

5. On critical case studies, see Eckstein (1975).

6. The term "mutual adjustments" was first used by Lindblom (1968) to describe informal cooperative efforts among proximate policymakers to arrive at an agreement on policy. Barzelay (1986) expanded it to study adjustments between private investors and the state.

7. Przeworski and Teune (1970).

8. Argentina had recurrent military interventions between 1955 and 1983, and Brazil was continuously ruled by a military regime between 1964 and 1985.

9. On the importance of studying sectors rather than sweeping aggregate national patterns, see Kitschelt (1991).

10. The emphasis on exports as a means to address foreign exchange shortages did not cancel out import-substituting efforts in consumer durables and capital goods (Kaufman 1990:129).

11. Canak (1984); Klapp (1987); Trebat (1983).

12. By the early 1970s Brazil was the second largest producer of capital goods in the developing world, after China.

13. For explanations focusing on a sector's position in the world economy, see also Gourevitch (1986).

14. Evans (1979) and Samuels (1987) argue that states are drawn into ownership where they *cannot* protect domestic producers. Feigenbaum (1985) also emphasizes the objective of competing in the international market.

15. On "state subsidiarity" in Latin America as limiting state expansion in the economy to its minimal expression (only as a last resort), see Stepan (1978) and Rouquié (1982).

16. On nuclear exports by Brazil and Argentina, see Solingen (1990a and 1991) and Tanis and Ramberg (1990).

17. On the organizational strength, political cohesiveness, and industrial leadership of this sector, see Boschi (1979 : 129).

18. Cardoso (1975) and Evans (1979), among others, describe the political power and access to state bureaucracies enjoyed by Brazil's largest industrial groups.

19. The UIA was represented in the most powerful umbrella entrepreneurial association, ACIEL (Imaz 1964; W. Smith 1991).

20. If only the economic component is taken into account, state A is presumably a more powerful state than B if A commands a greater share of global trade and credit and is less dependent on the global economy, as measured by the size of its external sector relative to GNP (Krasner 1976). For an analysis of Brazil's and Argentina's structural power, see Mares (1988).

21. Brazil became the world's tenth largest industrial producer in the 1970s; its percentage of global GNP (1.6 percent) more than doubled that of Argentina (0.6 percent). See *Economic and Social Progress in Latin America*, Inter-American Development Bank (1988 Report). On the measurement of Argentine decline, see Waisman (1987) and W. Smith (1991).

22. Adler (1987) and Sikkink (1991) provide two applications of this brand of theory to industrial policy in Brazil and Argentina. Adler focuses mainly on informatics and traces outcomes (policy regarding foreign investment) to the ideology of self-reliance of technocrats. Studies by Odell (1982), Ruggie (1982), Feigenbaum (1985), and Sikkink (1991) provide a more nuanced understanding of the independent power of ideas on political outcomes, by internalizing the domestic and international context in which ideas can make a difference.

23. Evans (1986) points out in his analysis of informatics in Brazil that the power of the technocrats—which shared a certain nationalistic ideology—should not be exaggerated and the agency's policies "were shaped by more than its personnel."

24. See Odell (1982), Hall (1989), Weir (1989), and Gourevitch (1989) for an analysis of the utility and limitations of cognitive approaches.

25. In that sense, this study bears an affinity with Grieco's (1984) analysis

of informatics in India and its emphasis on institutional interests, state autonomy, and state-private sector competition and to Haggard's (1990) comprehensive comparison of the politics of growth among NICs, as well as to Kitschelt's (1982) explanation of state intervention in nuclear industries in industrialized countries.

26. Feigenbaum (1985:17).

27. Haggard (1990); March and Olsen (1989); Hart (1992).

28. Zysman (1977) and Krasner (1978) highlight the influence of institutions and procedures on policy outcomes. Ikenberry et al. (1988:223) emphasize the influence of organizational features of the state—which are relatively resilient against the idiosyncratic actions of groups and individuals—on state preferences.

29. The original conceptualization of macropolitical consensus and lateral autonomy appears in Solingen (1993).

30. See Barzelay (1986:100) and H. Simon (1976:63). On the hierarchy and identification of goals and means—objectives and instruments—see Katzenstein (1978).

31. Preferences articulate primarily material, but also ideal, interests (Gourevitch 1989).

32. Private groups may be part of a coalition while being excluded from actual policymaking and implementation (Johnson 1982).

33. On "historic blocs," see Cox (1987); on homogeneous politico-economic "historical blocs," see Gramsci (1988:220). On size requirements of ruling coalitions in Latin America, see Ames (1987:34–40).

34. See Eckstein (1966) on extensiveness and intensity of political divisions.

35. For an example of such "moving equilibrium" in Thailand, see Doner (1992).

36. See, for instance, Katzenstein's (1978) comparison among the U.S., Britain, Japan, West Germany, and France. For an analysis of the coherence of macropolitical goals in France over time, see Milner (1987).

37. This characterization of Japan is from Pempel (1982:3).

38. On Japan's grand coalition, see Okimoto (1989). On Japanese consensus and its limits, see Katzenstein (1978), Johnson (1982), Borrus and Zysman (1986), Calder (1988), Krauss (1992), and Hart (1992). On the postwar consensus in Germany—low acceptance of inflation, tight money, fiscal frugality—see Vernon and Spar (1989). On East Asian NICs, see Haggard (1990). On India's consensus and deviations from it, see Encarnation (1989:220–21). On smaller European states, see Eckstein (1966), Katzenstein (1984, 1985), and March and Olsen (1989).

39. A sample of this debate includes Zysman and Tyson (1983), Nye (1990), Rosecrance (1990), and Nau (1990).

40. See Deyo (1987), Haggard (1990), Gereffi and Wyman (1990), and Onis (1991).

41. State autonomy refers to the ability to act independently from the

power of social classes or interest groups. It does not preclude high convergence of interests among state and private interests. It is a highly contingent or historically specific phenomenon rather than an absolute condition and can vary across countries and periods (Katzenstein 1978; Skocpol 1985; Klapp 1987; Migdal 1988; Onis 1991).

42. On the impact of a centralized, coherent state apparatus on East Asian NICs, see also Haggard (1990:45).

43. Deyo (1987:230–32). High levels of class conflict can contribute to strengthen consensus among otherwise adversarial industrial groups, which converge to restore confidence and stability (Frieden 1991).

44. Solinger (1991:27–30) emphasizes perceived economic threats as one foundation for elite consensus in China in the late 1970s. See also Zysman (1977) on France and Johnson (1987:145) on Japan. However, crises can also weaken consensus (Gourevitch 1986:21–34), exacerbate low levels of consensus (as in Argentina since 1955), or replace one kind of consensus for another.

45. Gourevitch (1986).

46. On conflicting objectives and bureaucratic bargaining, see Barzelay (1986) and Bendor and Hammond (1992).

47. The principle that "goal ambiguity" reduces central influence comes from organizational theory. See Cyert and March 1963) and Downs (1967).

48. On how strong central agencies increase the coherence of sector-specific policies, see also Katzenstein (1978). On the origins of hegemonic finance ministries as key loci of adjustment of domestic and international policy, see Cox (1986). On the British Treasury, see Weir (1989). On the limits of MITI's hegemony, and on growing bureaucratic segmentation and intra-bureaucratic conflict in Japan, see Samuels (1987), Calder (1988), and Hart (1992).

49. In Haggard's words, "'pieces' of an internally divided state are more likely to seek societal clients" (1990:45).

50. Haggard and Cheng (1987:45); Johnson (1987). On consensus and bureaucratic homogeneity in France, see Wilsford (1989).

51. Allison (1971).   52. Barzelay (1986).

53. Wilson (1989:183).   54. Tullock (1965).

55. On bureaucratic insulation as autonomy from private clientelistic pressures, see Schwartzman (1988), Geddes (1990), and B. Schneider (1993).

56. Barzelay (1986) uses the term "segmentation" to identify a decision-making arena where authority over the same political product is diffused among numerous collegial state agencies. The literal opposite of autonomy is "heteronomy" (subordination); however, perhaps the best logical opposite of the term in this context would be "synonomy" (together with)—a nonexistent word. I forgo the temptation to use this term and adopt the less specific, but more widely used, concept of segmentation. I thank Harry Eckstein for helping me settle these terminological issues.

57. B. Schneider (1991).

58. Encarnation and Wells (1985); Niskanen (1971). This is often an important source of autonomy for highly technical agencies, like atomic energy commissions.

59. See Arian (1989), Sprinzak and Diamond (1993), and *Ma'ariv*, June 28, 1992, on Israel and Lewis et al. (1991) on China.

60. On how agencies that monopolized the supply of a public service have weakened presidential control over those programs in the United States, see Niskanen (1971:196).

61. See Katzenstein (1978), Zysman (1977), Campbell (1988), and Ikenberry et al. (1988).

62. On CNEA's control over the nuclear program and the inability of other agencies, including the Foreign Ministry, to influence nuclear policy, see *La Prensa*, Mar. 22, 1984, p. 7.

63. The idiosyncratic characteristics of an agency relate to what Halperin (1974) labels "organizational essence."

64. These areas of intrabureaucratic interdependence are abstracted from Oszlak (1986).

65. On grand strategy, see Kennedy (1987:350). Calder (1988) discusses the impact of overlapping bureaucratic jurisdictions on Japan's foreign policy in high-tech areas, which he characterizes as erratic and reactive. On the centrifugal tendencies of state agencies, see Migdal (1988). On the consequences of a highly segmented policy process in the French oil sector, see Feigenbaum (1985).

66. Braybrooke and Lindblom (1963:130, 138).

67. See the examples in note 65.

68. High consensus can even dilute the preferences of state agencies like atomic energy commissions, which otherwise are often able to increase their autonomy from other state agencies by virtue of their built-in technological imperative.

69. Such convergence is evident, for instance, in the case of Singapore, where the Economic Development Board concentrated full authority over foreign investment (free from the regular ministerial structure) in the midst of a highly consensual industrial strategy (Encarnation and Wells 1985:61−62). On the notion of happy convergence, as it applies to state and private actors, see Jacobsen and Hofhansel (1984:197).

70. The term is from Deyo (1987).

71. A win-set is the set of all possible international agreements that would be acceptable to the relevant domestic constituents (Putnam 1988:437).

72. On this point the approach shares similar assumptions with studies of nuclear industries by Kitschelt (1986) and Jasper (1990).

73. On the reciprocal influence between technology and political structures, see Rycroft and Szyliowicz (1980), Piore and Sabel (1984), Kitschelt (1991), and Hart (1992).

74. General studies of the Brazilian and Argentine nuclear programs include articles and book chapters by Rosenbaum (1975), Quester (1979),

Courtney (1980), Redick (1978, 1981), Poneman (1982), Myers (1984), Lima (1986), and Adler (1987), mostly oriented to assess their respective political incentives to develop a nuclear program. Soares (1974) and Watson (1984) are unpublished doctoral dissertations wholly devoted to the Brazilian and Argentine cases, respectively.

75. A working definition of technology understands it as knowledge (embodied in equipment and machinery and disembodied in blueprints and licenses) based on scientific and technical criteria that can be applied to productive activities by improving or creating new goods and services (Dunning 1982).

76. On the advantages of research designs that combine cross-spatial and cross-time variance and permit "slides of synchronic comparisons through time," see Bartolini (1993).

77. Thus, the study of a nuclear industry provides us with a "crucial case study" in yet another sense. External constraints are assumed to be stronger where the bargain is over high technology. If such prediction fails in a "most likely case" such as a nuclear industry, theoretical propositions regarding the relative bargaining strength of technology recipients and suppliers need to be amended or discarded accordingly.

Chapter Two

1. In India, a state-owned enterprise (Bharat Heavy Electricals—BHEL)— India's largest engineering company—supplied heavy power equipment for the nuclear industry (Surrey 1987).

2. Most of these components are part of the nuclear system (as opposed to the conventional portions of a reactor) and require, therefore, more sophisticated industrial-technical capabilities.

3. CONUAR's majority partner was the private group Pecom; the firm was designed to produce nuclear fuel pellets.

4. On Brazil's capital goods sector as the most advanced in Latin America, see Gereffi (1990:91). On Argentina, see Díaz-Alejandro (1970:526–29, 532), and Dorfman (1983).

5. See IPEA (1984), Rosa (1984), Marton (1986:126–27), Katz (1985:133), and Rouquié (1987:291).

6. Study performed by Bechtel in 1974 for the Brazilian Nuclear Technology Corporation—a predecessor of Nuclebrás—created in 1971 (*Relatório* 1984). An earlier, aborted proposal under President Quadros in 1961 had also estimated a local capacity to supply 70 percent of inputs (*Manchete*, Apr. 24, 1976, pp. 75–97).

7. Only the steam generator, the reactor core, the reactor vessel, and the pressurizer could not have been produced by domestic private firms, according to a Congressional deposition by industrialist Claudio Bardella (*Relatório* 6, no. 5, p. 223). On capabilities in the engineering area, see Muller et al. (1980) and an interview with a former director of NUCLEN (Nuclebrás En-

gineering) in *O Estado de São Paulo*, June 7, 1985. See also deposition by industrialist Ramon Villares (*Relatório* 6, no. 5, p. 300).

8. Inputs supplied by private Brazilian firms would grow slowly, well into the future, to about 40 percent and 45 percent of total supplies for plants four through seven (*Relatório* 6, no. 5, p. 248).

9. See declaration by ABDIB president Silvio A. Puppo (*O Estado de São Paulo*, Dec. 19, 1978). A Nuclebrás official acknowledged the sector's capability in this area and justified its exclusion for commercial considerations (*O Estado de São Paulo*, Mar. 19, 1977, p. 29). See also deposition by industrialist Carlos Villares, who contended that given the preference granted to KWU in the supply of turbogenerators and pumps, his firm's effective participation was doomed (*Relatório* 6, no. 5, p. 302). Argentina's domestic capacity in this area did not even allow it to seek national participation in the supply of turbogenerators. In fact, even for the second plant, Argentine participation was no higher than about 6 percent (of the total in that area) (Tanis 1985).

10. Tanis (1985, 1986).

11. Unpackaging refers to the attempt to separate a project into its technological components to maximize national participation (Sagasti 1981a: 66–67).

12. On state autonomy in Brazil, see Schmitter (1971) and Stepan (1985). State autonomy was less challenged between 1964 and 1974 than, for instance, since the late 1970s. According to Deyo (1987), state autonomy in both Brazil and Argentina was historically lower than in South Korea or Taiwan, where there were no entrenched agrarian or comprador commercial classes.

13. On the merits and limits of a Brumairean explanation of state autonomy in Brazil—the bourgeoisie exchanges the right to rule for state protection of its basic interests—see Cardoso (1973), Quijano (1974), Kaplan (1974), and Faucher (1981).

14. On the political interests of these economic forces, see Martins (1977) and Frieden (1991).

15. Evans (1979).

16. Collier (1979), Sheahan (1987), Kaufman (1990), and Packenham (1992) highlight Brazil's broad consensus on economic strategy, in sharp contrast to Argentina. On the sources of political cleavage and consensus in Brazil, see also McDonough (1981). Brazil's relatively strong consensus did not include popular sectors (Ames 1987:14, 93–94); this was an exclusionary regime with a record of highly repressive methods toward working classes, aimed at imposing the model onto the rest of society.

17. Kaufman (1990:129).

18. See Evans (1979), Kaufman (1979), Ames (1987), Haggard (1990), and Frieden (1991). The argument relates to the initial definition of the program through the 1975 agreements. I explore the impact of changes in consensus in the latter part of the 1970s later in this chapter.

19. The 1973 increase in oil prices transformed a $7 million balance of payments surplus in 1973 into a $4.26 billion deficit in 1974. By 1974 the country had to set aside almost half of its export earnings to pay for imported oil. The foreign debt, $10 billion in 1972, grew to $22 billion in 1975. At the same time, electricity consumption was growing at a rate of over 13 percent per annum, and, at least according to some estimates at the time, the prospects for hydroelectrical and other sources of energy were quite bleak. On Brazil's energy balance, see Erickson (1981b) and Solingen (1991).

20. On the 1969–73 model and its aftermath, see Fishlow (1989) and Haggard (1990). On the nuclear program as an important piece in the strategy to eradicate energy as a serious bottleneck in Brazil's industrialization, see Vargas (1976:1032) and Soares (1976:213).

21. Barros (1978:132) supports the general claim that overall state-led economic growth took priority over strengthening the private sector. The number of state enterprises increased from 81 in 1959 to 251 in 1980 (Trebat 1983). In fact, just between 1970 and 1975, more state enterprises were created than during the preceding 30 years combined (Domínguez 1987:256). The specific contention that decision makers regarded fast implementation of a nuclear program as far more critical than maximizing private sector supplies is clear from a letter by Nuclebrás's director, Paulo Nogueira Batista, to the firm Bardella, urging it to make an immediate decision over their participation (*Relatório* 6, no. 5, p. 246).

22. On state entrepreneurship in Brazil, see Evans (1979); Hewlett (1980); Villela (1984).

23. On the potential role of nuclear energy in the steel sector, see Soares (1976:213).

24. Foreign borrowing in the mid-1970s enabled the regime to pursue an expansionist macroeconomic policy without resorting to a politically costly stabilization program (Haggard 1990; Frieden 1991).

25. On the centralization of economic policy-making in Brazil between the late 1960s and the mid-1970s, see Schmitter (1973), Diniz Cerqueira and Boschi (1977), Barros (1978), Boschi (1979), and Evans (1979).

26. Abranches (1978:149–50).

27. See inquiry by a parliamentary committee on the Nuclear Program (*Relatório* 3, p. 103).

28. On the powerful Ministry of Planning (particularly under President Castello Branco) and its responsibility to plan Brazil's economic affairs synoptically, see Barzelay (1986). On the hegemonic position of the Ministry of Finance within the bureaucracy between 1967 and 1974, with the ascendancy of Delfim Neto as Finance Minister, see Abranches (1978) and Kaufman (1979). On the constraining role of central economic agencies over sectoral programs during that period, see also Abranches (1978), Góes (1978), L. Martins (1986), and B. Schneider (1991). On the subsidiary role of the Foreign Ministry (Itamaraty) on nuclear energy decisions, see R. Schneider (1976:91).

29. On the proliferation of agencies with overlapping jurisdictions in Bra-

zil, see Stepan (1971:245). On overlapping jurisdictions over state enterprises in the energy sector, see B. Schneider (1987:555).

30. Two key decision makers in the nuclear agreements, President Geisel
and Energy Minister Ueki, were both associated with a tradition of state entrepreneurship, as former directors of Petrobrás, the state's oil monopoly. In
fact, Petrobrás was the model according to which Nuclebrás was organized
in 1974 (Biasi 1979).

31. See statement by Minister of Mines and Energy Shigeaki Ueki in *Manchete*, Apr. 24, 1976, pp. 75–97.

32. See B. Schneider (1987). On the role of technocrats, see Collier (1979).
The military never challenged core economic or other major public policy,
including nuclear policy, according to L. Martins (1986:225) and Bresser Pereira (1978). As individuals, however, President Geisel and (National Security Adviser) Golbery played a central role in defining and implementing Brazil's general economic strategy in those years (Abranches 1978:149) and the
nuclear program in particular (Skidmore 1988). Private interviews (conducted in São Paulo in May 1985) confirmed the limited institutional role
played by the military's National Security Council in the nuclear decisions
of 1974–75. This view is compatible with attempts by President Geisel to
reduce the power of the military High Command, as described in Stepan
(1988).

33. On the SNI, see Abranches (1978:120), Barros (1978:43–48), Dreifuss
(1981), Trebat (1983), Alves (1983), and Stepan (1988).

34. Dye and Silva (1979) argue that too much emphasis on the preemi-
nence of the military by students of the post-1964 regime obscured the economic character of what they call the "national project."

35. Skidmore (1988:194–96). Disappointments with the pace of implementation and with flaws in the program's technology transfer mechanisms
led the military in the late 1970s to promote a so-called parallel program.

36. See Kaufman (1979), Cavarozzi (1986), Waisman (1987), Rock (1987),
Stepan (1988), O'Donnell (1988), W. Smith (1991), Calvert and Calvert (1989),
Di Tella and Dornbusch (1989), P. Lewis (1990), Sikkink (1991), Kaufman
(1990), and Erro (1993). The armed forces ousted President Juan M. Perón in
1955, marking the beginning of a cycle of military juntas and brief constitutional interludes.

37. See O'Donnell (1973, 1976), Cardoso (1978), Sábato and Schvarzer
(1983), Rouquié (1987), W. Smith (1991), and Frieden (1991). On the tensions and convergence between agroexporting and industrial interests, see
also Turner (1983).

38. These contradictions are evident from statements, open letters, and
annual reports by the agroexporting association Sociedad Rural, the Unión
Industrial Argentina, and the Argentina Chamber of Commerce (W. Smith,
1991:84–90).

39. On the failure of powerful economic groups to articulate a common
program, see W. Smith (1985).

40. Erro (1993).

41. On the mass urban uprising of May 1969 known as the Cordobazo, see W. Smith (1985:42). On the exclusion of labor, see Kaufman (1979) and Collier and Collier (1991).

42. This was particularly evident in the episodes of 1955–57, 1962–63, 1966–73, and 1976 (Schvarzer 1983:121).

43. Kaufman (1979).

44. Erro (1993).

45. Between 1941 and 1970, ministers of the economy averaged only .88 years (321 days) each in power. There were years (1955, 1962, 1963) when three different ministers occupied that position, often replaced by individuals representing divergent interests and outlooks (Most 1991:56).

46. On failed efforts by Minister of the Economy Martínez de Hoz to control sectoral deviances, see Schvarzer (1983) and Pion-Berlin (1989:57–60). On balkanization of the state apparatus, see W. Smith (1991:176).

47. Oszlak (1984) has labeled this compartmentalization of spheres of influence "the feudalization of the state apparatus." Fontana's (1987) analysis of this "corporate pact" in Argentina traces most internal struggles regarding policy or procedure to intraservice rivalries.

48. Imaz (1964); Goldwert (1972); Potash (1980).

49. Mallon and Sorrouille (1975); Randall (1978); Potash (1980).

50. Poneman (1982); *Latinamerica Press,* Mar. 6, 1986. DGFM also absorbed up to 7 percent of the national budget annually (P. Lewis (1990:451).

51. Rouquié (1982:82).

52. Oszlak (1976). In contrast to the other two services, there was remarkable stability in Navy leadership as well, particularly between 1955 and 1968.

53. Argentina's dependence on imported energy was minimal, although a rich hydroelectrical and fossil fuel infrastructure was never efficiently tapped and left the country chronically subject to shortages (Watson 1990). On agents and principals in institutions, see North (1981).

54. In a conflict over the direction of their latest coup, one faction (the Azules, mostly from the Army's cavalry, with support from the Air Force) confronted the Colorados (who mustered the Navy and the Army's infantry and engineer corps). The Azules wanted to reduce the power of the Navy (Springer 1968:145–49). In the aftermath of the Colorados's defeat—following the use of armored units and naval air counterattacks—the Azules went ahead with their planned elections for 1963. The Navy's defeat allowed the victors to force a reduction in the number of admirals from 27 to 2, and of naval marines from 8,000 to 2,500. For an account of intramilitary cleavages, see Goldwert (1972), Rouquié (1982), Wynia (1986), and O'Donnell (1988).

55. The strategy thus attracted nationalists of various persuasions, including Peronists (quite a paradox for an institution controlled by the Navy, which had spearheaded Perón's dismissal).

56. On the early phases of CNEA, see Poneman (1984) and Mariscotti (1985).

57. Imaz (1964); Wynia (1986). On the Navy's traditional suspicion of extreme right-wing nationalist-statist currents in the Army, see Potash (1980: 218–21). Goldwert (1972) and O'Donnell (1988) analyze interest and ideological cleavages within the armed forces.

58. Poneman (1984:871); *Energeia* 44 (Mar. 1984):59.

59. Rouquié (1982); Sábato and Schvarzer (1983); Fontana (1987); Nogués (1986).

60. Perón provided lavish support for DGFM and established DINFIA, DINIE, and AFNE in the early 1950s. DINIE was created to run 38 formerly German firms, primarily in the pharmaceutical, chemical, electrical, and construction sectors, which were taken over as enemy property at the end of World War II (Randall 1978; Most 1991).

61. W. Smith (1991).

62. Oliveira (1976); Coelho (1976); Barros (1978:43–48 and 210); Ames (1987); P. Lewis (1990). Socialization processes and purges strengthened professional ideological homogeneity (Stepan 1973b).

63. Skidmore (1973:17).

64. See statement by former Nuclebrás director Paulo Nogueira Batista in *O Estado de São Paulo*, Oct. 19, 1983. Light water reactors—fueled by enriched uranium—use regular water as coolant and moderator, whereas reactors fueled by natural uranium use heavy water. Pressurized water reactors, a type of light water reactor, are more technologically advanced and economically proven. Out of 582 commercial and research reactors operating worldwide by 1982, 52 percent were of the pressurized type, as opposed to only 10 percent relying on heavy water (*Relatório* 5, p. 73).

65. The development of a "national" reactor could have taken 20–25 years, from the conceptual phase to the industrial phase (Biasi 1979:88, *Relatório* 6, no. 5, p. 157, and 6, no. 1, p. 311). The need to master reprocessing and enrichment would have arguably compounded the difficulties of designing a Brazilian reactor in a reasonably short time frame.

66. Most European countries, including France, Britain, and Sweden, initially embraced heavy water cycles, until cost and technical considerations persuaded them to shift to light water reactors (Jasper 1990). Natural uranium reactors are particularly suitable to produce plutonium, the stuff of nuclear weapons.

67. Rosa (1978).

68. Poneman (1982).

69. Sábato (1973).

70. *Foreign Broadcast Information Service, Daily Report* (Mar. 16, 1973).

71. The Navy opposed the policies of Martínez de Hoz and pressured its partners in the military junta to replace him. Martínez de Hoz was regarded as a threat to all military enterprises. Confrontations between President-General Videla and Navy Chief Admiral Massera personalized the institutional friction between the Army and the Navy at this time (Fontana 1987: 57–68).

72. Following the authoritarian years of Vargas's Estado Novo (1937–45),

there was a rather lengthy democratic interlude between 1946 and 1964. On the 1930–64 period, see Geddes (1994).

73. Attempts by each president since the early 1950s to formulate a coherent economic and industrial policy had failed to gain the support of a winning political coalition, although President Kubitschek came close to consolidating a consensual development strategy (Stepan 1978; Sikkink 1991). See also Frieden (1991).

74. Skidmore (1973). The middle and upper classes feared a "subversive" threat from the left, leading to profound socioeconomic transformations. Militant and radical labor leaders supported Goulart (Erickson 1977; Y. Cohen 1989).

75. Physicist Marcelo Damy was then Director of CNEN. The leading research programs pursued these technical options at the São Paulo Instituto de Energia Atomica and the Instituto de Pesquisas Radiológicas in Belo Horizonte.

76. National participation was estimated to reach close to 70 percent (*Manchete*, Apr. 24, 1976, pp. 75–97).

77. Meneses and Simon (1981).

78. Early discussions in 1967 were conducted by ten experts from the Nuclear Energy Commission, the utility Eletrobrás, the Ministry of Mines and Energy, and the National Security Council (*Relatório* 6, no. 5, p. 373, and 6, no. 6, p. 53). On the decision-making process for Angra 1, see Biasi (1979) and *Manchete* (Nov. 5, 1986).

79. See Stepan (1985), Barzelay (1986), and Evans (1979) on the external sources of erosion of consensus. Fishlow (1989) argues that "frantic inconsistency" had become the hallmark of public policy by the early 1980s.

80. On the *Documento dos Oito*, see Payne (1994:64–65).

81. In 1986 CNEN became, once again, directly subordinated to Brazil's presidency and to the National Security Council (*Folha de São Paulo*, Nov. 11, 1986, p. 16). Unlike CNEA, CNEN had been administered, since its inception in 1956, by a variety of generals, admirals, and scientists. By the 1980s, however, CNEN maintained strong links to the Navy, including budgetary support (*Tendencia*, Mar. 1984, pp. 40–42). CNEN came to control the secret so-called parallel program and to promote national technologies (*Folha de São Paulo*, Apr. 28, 1985, p. 25).

82. Evidence for these claims can be deduced from failed attempted challenges to CNEA along these lines (Poneman 1982) and from the general risk-averse behavior of utilities regarding technological options and reactor types (Kitschelt 1991).

83. Rouquié (1987:299).

84. On Argentina's high level of consensus under Menem, unprecedented since the early 1950s, see Packenham (1994).

85. Hollis and Smith (1990:146). Structures suggest, as Strange (1983) argues, a "supporting framework.".

86. On the development of supporting coalitions as an instrument to ac-

cumulate political influence, see Bates (1981), Ames (1987), and Geddes (1994).

87. For a detailed analysis of the acerbic distributional struggle among the three services and the Navy's attempt to retain its slipping position, see Fontana (1987:42–90).

88. Such path-dependent scenarios weaken even further the ability to infer with certainty where an agency stands from "where it sits," ignoring the historical context and political evolution at great peril. On path-dependent sequences in economic change, see David (1985) and North (1990). On the development of institutional routines, see Krasner (1984), Ikenberry (1988), and March and Olsen (1989).

89. The fact that the choices of the Ministries of Planning and Finance in Brazil changed over time in cyclical fashion is further proof that choices are not easily derived from institutional position and may be highly responsive to political changes in the ruling coalition.

Chapter Three

1. For a comprehensive analysis of Brazilian industrialists and their evolving political preferences, see Diniz and Boschi (1978) and Boschi (1979).

2. On the "triple alliance," see Bresser Pereira (1978), Boschi (1979), and Evans (1979).

3. According to Evans (1982), strengthening this segment of the industrial bourgeoisie was a strategic objective in maintaining capitalist industrialization. On the initial support of capital goods producers for President Geisel and their eventual disillusionment, see Ames (1987:192).

4. Only 30 percent of the equipment for nuclear reactors is strictly nuclear. The rest is supplied by conventional manufacturers, mostly in the areas of mechanical and electrical engineering. Capital goods in Brazil's industrial census classification include sections of the metallurgical, mechanical, electrical, and transportation equipment industries (excluding passenger automobiles and consumer durables). Capital goods account for the bulk of the mechanical industry's production and over half of the electrical industry (Erber 1977:2).

5. Villela (1984). Leff (1968) suggested, from a strictly economic (domestic capital formation) viewpoint, that Brazilian policymakers should look elsewhere for a stimulus to rapid economic growth. Others, such as Hirschman (1958), Erber (1977, 1984), and ECLA studies, brought to the forefront the critical role assigned to capital goods in developing countries (given foreign exchange constraints) and the dynamic effects of "vertical interdependence" among producers of consumer and intermediary goods and capital goods industries. On the development of capital goods and the automobile sector in Brazil, see Kurth (1979).

6. Wasserman et al. (1976).

7. Value added grew from 1949 to 1959 by 10.3 percent annually, reach-

ing about 33 percent of value added in the manufacturing sector (Leff 1968:167).

8. Erber (1977:5) and (1984:12).

9. Kaufman (1979:231).

10. Capital goods began receiving higher rates of protection than intermediate or consumer goods by 1974, in a reversal of the structure of protection applied since 1967.

11. Fishlow (1989). Between 1973 and 1977 import substitution in capital goods increased national production by a factor of three (Hewlett 1978). Local production accounted for 72 percent of consumption in 1970 and increased to 78 percent in 1979 (Chudnovsky and Nagao 1983:98). In the area of custom-made goods, where many nuclear components fall, local production accounted for less than 60 percent in the 1970s.

12. President Geisel approved an interministerial instruction to cooperate with CACEX in directing the maximum possible volume of orders (machinery, equipment, and engineering) to private Brazilian firms. Foreign firms' shares fell from 60 percent to 30 percent of total equity in some areas (World Bank Country Study, p. 107).

13. Fishlow (1989:93).

14. Leff (1968:140).

15. A substantial amount of private investment was financed from public resources such as the development bank (BNDE) and the National Housing Bank (BNH) (Baer 1983).

16. Fishlow (1989).

17. Almeida (1983:44).

18. On multinational corporations as a key pillar in the "triple alliance," see Evans (1979). On the dangers of domestic "deepening" for MNCs, see Serra (1979:122).

19. The German Deutsche Bundesrepublik Bank and a private consortium of 30 banks headed by Dresdner Bank provided financing for the nuclear program, contingent on the provision of German equipment and services (*Jornal do Brasil* June 27, 1976, p. 2).

20. The impact of these perceptions was compounded by the technological complexity and the "system-quality" (Hirschman 1967:43) of the nuclear program, but they reflected a more general pattern. In September 1975 (three months after the signing of the Nuclear Agreement and Commercial Directives on commercial contracts), the Secretariat of Industrial Technology (Ministry of Industry and Trade) organized a meeting on "Technology in the capital goods sector." There, private entrepreneurs voiced strong complaints over the resistance by state enterprises to local designs and over the preference of state enterprises for foreign licensing (Erber 1977:293).

21. On price considerations regarding inputs for nuclear plants, see *Jornal do Brasil* (June 5, 1976, p. 13).

22. Minister of Mines and Energy Shigeaki Ueki argued that local industry was not capable of supplying required materials and services, such as metal-

lic joints, special steels, and welding techniques (*Manchete*, Apr. 24, 1976, pp. 75–97).

23. The principal primary system components include the pressure vessel, heat exchangers, accumulators, and reactor structures (Lepecki 1985:31). See Appendix D.

24. Companhia Brasileira de Tecnologia Nuclear (1973). Only 34 systems (out of 1,312) were considered beyond the capability of Brazilian companies (pp. 1.8-2 and 2.1-6). These included the nuclear components: steam generator, reactor core and pressure vessel, and pressurizer. It should be noted that major nuclear suppliers (including KWU) also tended to import heavy components (such as pressure vessels) from Japan.

25. *Relatório* 4, p. 200. Professor R. A. Muller of Germany's Karlsruhe Nuclear Center expressed in 1978 that all the light and medium equipment necessary for the first plant could be produced by some thirteen Brazilian firms (*O Estado de São Paulo*, Mar. 15, 1978, p. 11).

26. ABDIB was created in 1955, when the import-substituting policies of Petrobrás gave considerable impetus to local production. ABDIB consistently criticized purchasing policies of most state enterprises, perhaps with the exception of Petrobrás and CESP (Erber 1977:292). By 1985 ABDIB's membership included roughly 60 percent wholly Brazilian firms, 40 percent foreign subsidiaries, and three state enterprises (NUCLEP, Usimec, and Mafersa). For a comprehensive analysis of ABDIB in its organizational and political dimensions, see Boschi (1979:181–222).

27. *Relatório* 6, no. 5, pp. 210 and 232. Projected investments by capital goods producers for 1975 were expected to increase the sector's capacity by 70 percent (*Relatório* 6, no. 5, p. 256).

28. See deposition by Claudio Bardella (*Relatório* 6, no. 5, pp. 256, 267, 276). In some instances, producers declined to supply a specific component for not being commercially attractive enough. At least some industrialists pointed to their lack of interest in a more extensive involvement, given unattractive returns (*Jornal do Brasil*, June 27, 1976, p. 2; *O Estado de São Paulo*, Sept. 20, 1979).

29. *Relatório* 6, no. 5, pp. 210 and 232. In March 1975, however, Nuclebrás had informed ABDIB of its intention to go ahead with the creation of a state enterprise in the area of heavy components.

30. *O Globo*, June 20, 1975, p. 32 and July 6, 1975, p. 7.

31. KWU (a subsidiary of Siemens) was allegedly eager to reserve production of the turbogenerator—equivalent to 30 percent of the cost of the plant—to its controlling firm. Newfarmer (1978) depicts an international market in electrical components characterized by restrictive commercial practices and entry barriers for potential competitors.

32. See declaration by ABDIB President Silvio A. Puppo (*O Estado de São Paulo*, Dec. 19, 1978). A Nuclebrás official acknowledged the sector's capability in this area and justified its exclusion on commercial grounds (*O Estado de São Paulo*, Mar. 19, 1977, p. 29). See also deposition by industrialist

Carlos Villares, who contended that given the preference granted to KWU in the supply of turbogenerators and pumps, his firm's effective participation was doomed (*Relatório 6*, no. 5, p. 302). His own proposal was based on a technology transfer agreement with the Swiss firm Brown-Bovery, a leading producer of turbogenerators.

33. *O Globo*, Nov. 24, 1975, p. 24; *Manchete*, Oct. 18, 1975, pp. 32–34.

34. The three firms were Bardella, Cobrasma, and Confab, with respective participations of 47 percent (Cobrasma), 36 percent (Confab), and 18 percent (Bardella) (*Relatório* 4, p. 205; *O Globo*, Jan. 14, 1976).

35. Personal interview with a former Nuclebrás official (Rio de Janeiro, May 19, 1985).

36. For an extremely useful approach to the study of private investment decisions in steered economies, see Barzelay (1986).

37. Growing opposition to Nuclebrás's objectives from other state agencies—Planning Secretariat, Eletrobrás (state utility holding company), BNDE, and others—placed heavy budgetary constraints on the program.

38. Minister Golbery de Couto e Silva reportedly conveyed the first estimate to private entrepreneurs directly (*O Globo*, Nov. 8, 1978). On the second estimate, see *Jornal do Brasil* (Nov. 7, 1976, p. 47).

39. KWU was involved in the selection of participating Brazilian firms at the outset, after surveying them with Brazilian officials in 1974.

40. On these general conditions, see Bresser Pereira (1978).

41. Hirschman (1979:68).

42. Critics of the nuclear program included the National Confederation of Industry, the Club of Engineers in Rio, and the Brazilian Bar Association, among others. Industrialist Ermírio de Moraes (president of the influential Votorantim group) even proposed postponement of the nuclear program in favor of less expensive hydroelectric alternatives (*Relatório 6*, no. 4, p. 20).

43. Stepan (1985). On the role of Brazil's industrialists in restoring the democratic process in Brazil in the 1970s, see Payne (1994).

44. Pressures mounted within the state bureaucracy as well and were reflected in the import-substitution policies of the Second National Development Plan supported by Planning Minister Reis Velloso, Industry and Trade Minister Severo Gomes, and National Economic Development Bank (BNDE) President Pereira Vianna (Anglade 1985:95). Gomes resigned following a rallying of orthodox forces that included private capitalists.

45. Stepan (1985). On the breakdown of the old "pact" between the bourgeoisie and state technocrats, see Bresser Pereira (1988).

46. Diniz and Lima, Jr. (1985:90).

47. *Jornal do Brasil*, Sept. 16, 1977, p. 21. The industrialists' use of the press in their campaign against foreign and state capital was not a new phenomenon (Leff 1968:112). However, their leverage and effectiveness were compounded by the widespread social and political discontent of the late 1970s.

48. Evans (1982).

49. By mid-1975, three months before signing the purchase contract for the first units, Brazilian firms had been allocated even lower shares in the supply of equipment, about 21 percent (*O Estado de São Paulo*, July 24, 1975, p. 19). On revised levels of participation, see Syllus (1982:20).

50. *Jornal do Brasil*, June 30, 1984.

51. A commitment of $1 billion over ten years was reported in *Manchete*, Apr. 24, 1976, pp. 75–97.

52. While *national* private firms accounted for about 30 percent of the mechanical sector's assets in 1972, their share increased to 53 percent in 1974 and to over 60 percent in 1979 (Baer et al. 1976:78). Brazilian firms controlled only 22 percent in electrical machinery by 1972, down from 39 percent in 1966 (Newfarmer and Mueller 1975). Baer (1983:193) reports multinationals as holding about 80 percent of Brazil's electrical industry in 1975. See also Boschi (1979:113). Some of the domestic components to be supplied for Angra 2 and 3 included: main cooling water pumps, steel containment, borated water storage tanks, heat exchangers, and cranes, as well as complete ventilation, air conditioning, and demineralized water systems.

53. *Brazil Energy*, Dec. 10, 1980, p. 5. See also *Relatório*, 6, no. 5, p. 269. By 1985, total orders amounted to $500 million at the most (Lepecki 1985: 30). According to *Brazil Energia* (Dec. 10, 1980, p. 5), capital goods firms received orders for $250 million (for the period 1980–86) for the first plant.

54. Layoffs in the capital goods sector—24,000 skilled workers in 1980–81—increased in 1982, as production fell at a yearly rate of 15 percent (*Brasil Energia*, Apr. 4, 1982, p. 13).

55. Four (out of eight) plants were canceled in 1982 (Plano 2000). By 1995, only the first plant planned under the 1975 agreements (Angra 2) was under construction and scheduled to begin operating at the end of the decade.

56. State firms dominated infrastructural areas (such as electricity generation, steel, oil, communications, and transportation) and competed with private firms, to some extent, in the petrochemical, mineral, banking, and financing sectors (Bresser Pereira 1978; Diniz and Boschi 1978; Barros 1978: 308–19).

57. *O Estado de São Paulo*, Dec. 17, 1978. Exports of heavy components were also foreseen, despite an impending crisis in the international reactor industry.

58. Nuclebrás officials also emphasized the importance of the geographical proximity of the heavy components factory to the ocean, to facilitate transportation to the plant site, a factor not compatible with the actual location of many of Brazil's heavy mechanical producers around São Paulo. Instead, the joint venture between Nuclebrás and German investors (NUCLEP) was to be located in a coastal locale, near Itaguai.

59. *Relatório* 6, no. 5, p. 295, and *Relatório* 3, p. 144.

60. These arguments were advanced by ABDIB's president, Silvio A. Puppo (*O Estado de São Paulo*, Dec. 19, 1978).

61. *O Estado de São Paulo*, Dec. 17, 1978. Heavy components for plants 1

and 3 were to be imported. NUCLEP was to produce such components for plants 3 through 8, but owing to the program's delays and uncertainties, that never happened, and those plants were ultimately canceled (*Gazeta Mercantil*, Dec. 27, 1984).

62. NUCLEP received an average of two requests per week, most of which were referred to the private sector, with the exception of two (for Petrobrás and Montreal Micoperi) (*Gazeta Mercantil*, Jan. 13, 1982).

63. *O Globo*, June 20, 1975, p. 32.

64. *O Estado de São Paulo*, Sept. 20, 1979.

65. Some Brazilians attributed this structure of technology transfer to German reluctance to diffuse technology to Brazilian firms, fearing international competition. German interests, in this view, may have been better served by transferring blueprints to NUCLEN, a joint venture where KWU exercised some control (private interview, Rio de Janeiro, Apr. 20, 1985). Rodolfo Andriani, president of a major engineering firm (Inepar), claimed it was close to impossible for Brazilian firms to win bids, given KWU's control of NUCLEN. His own firm had invested considerably in quality assurance systems (*O Estado de São Paulo*, Aug. 25, 1979, p. 3).

66. *Jornal do Brasil*, Jan. 25, 1977, p. 4; *Relatório* 4, p. 176.

67. *O Globo*, June 20, 1975, p. 32 and Apr. 25, 1976, p. 31.

68. Baer et al. (1976:11).

69. On this pattern of state intervention leading to related industrial activities, see Evans (1985:198).

70. Erber (1977:246).

71. Chapters 6 and 7 include a more extensive operational analysis of state promotion of technological capabilities in the private sector.

72. The interministerial CDI considered major private development projects, to assist them with fiscal incentives (Trebat 1983). The council could grant tariffs and tax exemptions to promote the use of domestically produced machinery and equipment. For a comprehensive analysis of CDI, see Abranches (1978:170–310).

73. Serra (1979); Villela and Baer (1980:190–92).

74. Almeida (1983).

75. *O Estado de São Paulo*, July 18, 1978. On the expanded power of the Planning and Finance Ministries in response to the 1973 oil crisis, see Stepan (1985:335).

76. *O Globo*, July 15, 1977, p. 23.

77. Abranches (1978:150).

78. Diniz and Boschi (1977:172); Boschi (1979:152); Diniz and Lima, Jr. (1985:48).

79. CACEX (Carteira de Comércio Exterior–Banco do Brasil), the Central Bank's foreign trade bureau, was generally responsible for ruling on the test of "similarity" (*Jornal do Brasil*, Nov. 30, 1977, p. 21).

80. *O Estado de São Paulo*, July 18, 1978. Capital goods producers criticized these presidential decrees, including one granting NUCLEI (Nuclebrás subsidiary in uranium enrichment) tax exemption for importation of (un-

usually overpriced) nuclear equipment (*O Estado de São Paulo*, Dec. 19, 1978).

81. Faucher (1980:21).

82. *Jornal do Brasil*, May 24, 1977, p. 25.

83. Hewlett (1980:55).

84. FINEP had had a program to support local consulting since 1973, providing finance for local consulting firms to acquire human and material resources and enabling them to substitute for imports (World Bank Country Study 1980:103). Another program supported technological development of private firms in priority areas of the National Development Program, financing R&D for new products and processes, adaptation and absorption of foreign technologies, and quality control measures.

85. *Manchete*, Apr. 24, 1976, pp. 75–97.

86. *Jornal do Brasil*, May 23, 1977, p. 13. BNDE's apparent antagonism toward Nuclebrás was reflected in a 1978 study, reported in *Jornal de Tecnologia e Ciencia* (1978–79). Surveying industrial capabilities in the area of custom-made mechanical equipment, the study was skeptical of technological gains to be accrued from contracts with German firms in the nuclear sector. BNDE maintained close relations with private entrepreneurs, at times acting as an intermediary between them and other state agencies (Diniz and Lima, Jr. 1985:48).

87. *Folha de São Paulo*, June 30, 1978. The BNH is oriented to civil construction programs.

88. Bardella, for instance, obtained 60 percent of its investments up to 1979 (over half a million dollars) from FINAME. The remaining 40 percent were self-financed (*Relatório 6*, no. 5, p. 293; *O Estado de São Paulo*, Jan. 15, 1977, p. 27). About $4 million was provided by FINAME in 1979–80 for the nuclear sector (*Relatório BNDE* 1980). Interestingly, the state firm NUCLEP received some support from BNDE's FINAME (*O Estado de São Paulo*, Dec. 17, 1978). The CDI also approved fiscal incentives for NUCLEP (*Folha de São Paulo*, Aug. 12, 1977, p. 22).

89. Interviews with private firms involved in the nuclear program. See Chapter 7.

90. Private interview, Rio de Janeiro (May 20, 1985).

91. *Jornal do Brasil*, May 5, 1976, p. 13. The costs of technology absorption and personnel training were deducted from the final price for local goods (*Relatório 6*, no. 5, p. 265; *O Estado de São Paulo*, Sept. 20, 1979).

92. Domestically produced machine tools in Brazil dropped from 64 percent of total consumption to 33 percent (between 1966 and 1970), while in Argentina the figures were 79 and 55 percent, respectively (Kaufman 1979:178).

93. W. Smith (1991).

94. Mallon and Sorrouille (1975:87). State enterprises were responsible for a significant percentage of these imports. Notably, enterprises associated with the Army's Fabricaciones Militares (in steel and petrochemicals) were traditionally oriented to foreign equipment and technology.

95. Tanis (1985); Sábato et al. (1978).

96. Tanis (1985). CNEA did make use of a variety of tax exemptions, relying on Law 18243, which was designed to strengthen Argentine products vis-à-vis foreign competitors. Such exemptions included sales tax, reimbursements according to decree 46/65, and exemption from import deposits and taxes on imported inputs not produced locally (Aráoz and Martínez-Vidal 1974:67).

97. Sábato (1973).

98. Tanis (1986). At first INVAP developed zirconium sponge (which had been previously imported), a material needed in the fabrication of fuel elements. It later provided services in instrumentation for a reactor exported to Peru, and in extractive metallurgy, electronic and optical instrumentation projects, special mechanical equipment, and the construction of an experimental reactor (RA-6) (*Clarín*, Nov. 21, 1983).

99. *Energía Nuclear*, vol. 14 (1983).

100. *Economic Information on Argentina*, Buenos Aires, 119, (1981):13.

101. For a comprehensive discussion of Argentina's corporate associations, see W. Smith (1991), who provides the main reference source for this section.

102. On how a firm's export orientation and its multinational operations define its preferences for free trade, see Milner (1988). On the specific views of different entrepreneurial sectors in Argentina and Brazil, see Domínguez (1982).

103. The CGE counted retired army general Juan Enrique Guglialmelli—director of a military program (CONADE) designed to "homogenize" and improve economic and technical decision making within the state apparatus—among its allies (W. Smith 1991). On the CGE, see also Imaz (1964) and Lewis (1990).

104. However, by the 1970s the UIA looked at protectionism more favorably than ACIEL as a whole and was prepared to join CGE industrialists in a common front (Lewis 1990:343).

105. W. Smith (1991).

106. Quoted in W. Smith (1991:170).

107. W. Smith (1985).

108. P. Lewis (1990:454). Military expenditures represented over 3 percent of the GDP in those years (Remmer 1989:195).

109. *El Desarrollo Nuclear Argentino*, p. 58.

110. Some of these firms were forced to dismiss over 14,000 employees in 1985 (*Clarín*, Apr. 17, 1985).

111. A few of these firms may not have been politically salient within their class organizations, but they held a publicly recognized technological leadership.

112. *Noticias Argentinas*, May 20, 1985.

113. On the sources of the shift toward a privatizing consensus in Argentina, see Packenham (1992).

114. CONUAR produced nuclear fuel with technology licensed from

CNEA (*Clarín*, June 25, 1991, p. 17), *FBIS* (Aug. 21, 1991, p. 5. On debates over privatization schemes, see *FBIS* (May 5, 1994, p. 13).

115. Nash (1991, 1992).

116. State enterprises in the electricity area owed CNEA $47.5 million in 1991 (*FBIS*, Aug. 21, 1991, p. 5), and an estimated $500 million in 1994 (*FBIS*, May 5, 1994, p. 13).

117. On the classic uncertainty of Argentina's investment climate, see Kaufman (1979:242–43), and Di Tella and Dornbusch (1989).

118. Although borrowing from Hirschman (1982), I use "shifting involvements" here in a different sense than originally specified in that work.

119. On these contradictions, see Barzelay (1986).

120. On this selectivity regarding local capital, see Bresser Pereira (1978).

121. A 30 percent participation in the provision of machinery and equipment for state projects was in the average range during the late 1960s and early 1970s.

122. Erber (1977:179).

123. On the absolute dominance of state enterprises within the largest corporations in Brazil, in contrast to India and South Korea in particular, see Encarnation (1989:194).

124. On this preemptive development of new state capacities to stem the power of transnational capital, see Evans (1985:195).

125. Faucher (1981:11–40).

126. L. Martins (1986:82). Cardoso (1975) describes the "*anéis burocráticos*" (bureaucratic rings) linking private entrepreneurs and state officials as ad hoc, particularistic, and intermittent, rather than stable and sectoral, and the state as playing the dominant role in all of them. For a converging view, see Diniz Cerqueira and Boschi (1977) and Boschi (1979).

127. *Relatório* 6, no. 5, p. 211. See also letter from industrialist Carlos Bardella to President Geisel in ibid., pp. 241 and 257. Bardella, a former president of ABDIB, had requested an interview with President Geisel as early as October 1974; the interview took place only in March 1976.

128. Barzelay (1986).

129. According to Munck (1984:225), the low number of manufacturers of heavy machinery in the early 1970s grew to 21 by 1978 (50 percent of them under national control). On the sources of growing political strength among Brazil's industrialists, see Payne (1994).

130. *O Estado de São Paulo*, Feb. 25, 1986, p. 2 and Dec. 18, 1986, p. 29. By 1983, CNEN was actively developing a Brazilian-designed research reactor, in cooperation with universities, research institutes, and private firms (*Folha de São Paulo*, Oct. 25, 1983, p. 8).

131. The enterprise would produce the 100 MWe PWR National Reactor (*O Globo*, May 17, 1991, p. 15). CNEN would be engaged only in R&D (*FBIS*, Sept. 12, 1991, p. 5).

132. Becker et al. (1987). As Becker suggests (1987:66), the local bourgeoisie exerts its power while combating and reducing foreign domination at the

same time that it uses its power to advance industrialization and expand its own influence and privilege.

133. Ibid., p. 54; Payne (1994).

134. For some, this process presaged the private sector's forsaking of its reliance on state power (Bresser Pereira 1984), whereas for others the private sector's support for state-controlled political and economic change continued (Cardoso 1986).

135. The phrase borrows from Hirschman (1971). On the antistatist campaign, see Stepan (1985).

136. On this mistrust, see Wynia (1978). On characterizations of Argentine entrepreneurs as "precapitalist" and short-term oriented, see Lewis (1990:329–35).

137. Becker et al. (1987:208).

138. Kuhn (1966:188).

139. Brazilian suppliers of the nuclear industry were generally more diversified than their Argentine counterparts; some firms favored hydroelectrical alternatives to a nuclear industry.

140. On the call for studying firms'—rather than industry's—preferences, see Milner (1988).

141. On how segmentation decreases the efficiency of political signals to private entrepreneurs in large and complex issue areas, see Barzelay (1986: 118). On private entrepreneurs' characterization of a highly segmented policy-making arena in capital goods more generally, see Boschi (1979: 162–70).

142. *Political* products are distinguished from *market* products "by the extent to which cash flows are determined by sector-specific policies rather than in the marketplace" (Barzelay 1986:57).

143. This proposition is adapted from Barzelay (1986:121).

Chapter Four

1. *O Estado de São Paulo*, June 10, 1984, p. 50; Dec. 21, 1985, p. 3; and Nov. 20, 1990, p. 16; *Financial Times*, Nov. 25, 1985.

2. On win-sets and involuntary defection, see Putnam (1988).

3. On the division of Argentine ministries among Argentina's three services, see Fontana (1987:33).

4. Gereffi (1983).

5. Stepan (1978) argues that a stronger, cohesive state elite—capable of providing a predictable environment for foreign capital—enjoys a better bargaining position.

6. On computers in India, see Grieco (1984). On automobiles in Mexico, see Bennett and Sharpe (1985). Encarnation and Wells (1985) analyze how segmented (or, in their term, "diffused") decision making increases the unpredictability of bargaining outcomes, therefore increasing the costs to the investor.

7. The "deal of the century" was really the product of a sequence of negotiating nodes. On October 31, 1974, Brazil signed the so-called Protocol of Brasília (Diretrizes para a Cooperação Industrial entre o Brasil e a Alemanha), a program of industrial cooperation in the nuclear sector between the governments of Brazil and West Germany. Both parties signed the Bonn Protocol (Cooperation Accord on the Peaceful Uses of Nuclear Energy) in June 1975 (to expire in 1995), including "Specific Directives" on cooperation and negotiation of commercial contracts that were incorporated in the Annexes. This amounted to the operationalization of the October 1974 agreements. On firm-level licensing and cooperation agreements between private German and Brazilian industrial firms, see Chapters 6 and 7.

8. Declarations by KWU officials at the time confirm that they regarded the 30 percent share of supplies by private Brazilian firms and the technological control of joint ventures by German firms as squarely within the domestic win-set (*Nucleonics Week*, Oct. 31, 1985, p. 6).

9. As argued, decision makers perceived the costs of a more gradual program—with higher levels of domestic participation—to be too high as well, because of their potential to disrupt immediate energy supply, a critical requirement of the model.

10. KWU's control of quality assurance qualifications in NUCLEN, as I will discuss shortly, was yet another means used by KWU to maximize its own supplies at the expense of Brazilian industry.

11. On the position of Brazilian negotiators, see declarations in *Manchete*, Apr. 4, 1976; *Journal do Brasil*, Nov. 7, 1976; *O Estado de São Paulo*, Mar. 19, 1977; *Relatório 6*, no. 5, p. 302; and *O Globo*, June 20, 1975.

12. The segmented nature of Brazil's negotiating team is evident from the nature of participants at different bargaining nodes: Minister Ueki went to West Germany in November 1974, Finance Minister Mario H. Simonsen in March 1975, FURNAS's President Luiz C. Magalhaes in April 1975, and Foreign Minister Azeredo de Silveira in June 1975—for the signing of the Bonn Protocol. Planning Minister João P. dos Reis Velloso and Energy Minister Shigeaki Ueki signed the commercial agreements with KWU in 1976 (*Relatório* 3, p. 78; *O Estado de São Paulo*, July 23, 1976, p. 35).

13. The policies of Finance Minister Delfim Netto had forced a search for foreign financing, even where it implied higher ratios of suppliers' credits—which often increased imports of equipment—over regular loans.

14. In reality, there was more happy convergence than conflict between Nuclebrás's leadership and central economic agencies, a fact that undermined Brazil's ability to use bureaucratic pressures as a bargaining chip. On dissenting voices within Nuclebrás, see Carvalho (1981), Carvalho and Goldemberg (1980), and Carvalho et al. (1987).

15. Negotiators can also be misinformed about levels of bureaucratic autonomy or segmentation. For an example of the costs of not knowing what agencies have the power to negotiate, see Encarnation and Wells (1985: 49, n. 8).

16. These technologies are considered sensitive because of their potential for nuclear weapons production.

17. Putnam (1988:447).

18. On state purchasing power as an instrument of technological development, see Caputo (1978), Erber (1977, 1984), and Hart (1992).

19. Sagasti (1981a:66–67).

20. This process is also referred to as unpackaging. "Black boxes" in technology transfer contexts imply the acquisition of a complete project (often through turnkey agreements), where the recipient is not necessarily required to understand the project's underlying conception but merely to be capable of operating the facility upon completion.

21. Sábato et al. (1978).

22. Dahlman and Westphal (1982:124). On the advantages of unbundling, see also Ruggie (1983).

23. Carvalho (1981).

24. The phases of a project include: (1) preinvestment, feasibility, and marketing studies (of the project's viability, on the basis of available information on input sources, choices of production technology, plant scales, and locations); (2) preliminary detailed studies of alternative technologies (using more specific engineering norms to select among alternatives on the basis of costs and mode of financing, timetable, and labor); (3) basic engineering (central processes embodied in flow sheets, layouts, performance specifications, and designs for major equipment and machinery; (4) detailed engineering (final design of mechanical and electrical systems, detailed architecture, construction, and civil engineering, and final equipment specifications); (5) procurement and construction (selection of suppliers of equipment, construction and assembly firms, bid evaluation, and coordination of various subcontracting activities and inspection); (6) training of personnel; and (7) start-up and initial troubleshooting services (overcoming various design problems). These phases are adapted from Dahlman and Westphal (1982).

25. Dahlman et al. (1985:31).

26. For a detailed description of nuclear technology transfer evaluation by CNEA officials, see Sábato (1973) and Tanis (1985:17–25).

27. This group included CNEA's president, the directors of technology, energy, and technical assistance to industry, and fifteen other CNEA professionals (Sábato et al. 1978).

28. Turnkey projects are considered to be quintessential black boxes, particularly if there is no transfer of fundamental information, or specific efforts to absorb it, or both (Dahlman et al. 1985:31). Yet Westphal et al. (1984) found that South Korea assimilated technological know-how mainly through learning-by-doing, using mostly direct foreign investment, machinery imports, and turnkey plant construction.

29. Failed attempted challenges to CNEA along these lines provide evidence for this claim (Sábato 1973; Poneman 1982). The utilities SEGBA and Agua y Electricidad (Water and Electricity) purchased electricity from

CNEA's Atucha 1 and Embalse and were naturally inclined to more rationalized prices.

30. CNEN had supported, in the late 1960s and early 1970s, the acquisition of a turnkey plant from Westinghouse, Brazil's first nuclear plant (Angra 1), negotiated mostly by technocrats from the electrical utilities, including Eletrobrás, Furnas, and Ministry of Mines and Energy, and the Interministerial Energy Commission (*Manchete*, Nov. 5, 1986).

31. See deposition by Jair Mello (*Relatório 6*, no. 5, p. 373, and *Relatório 3*, p. 103). Technical personnel and professionals were not entirely excluded from preliminary negotiations but did not play any role in the decisive ones (*Relatório 6*, no. 6, p. 53). According to physicist Rogerio Cerqueira Leite, Nuclebrás was not blessed with a leadership competent in the technical and scientific aspects of the nuclear agreement, and did not allow serious input of this nature into its decision-making structure (Leite 1978).

32. This agreement was thus defined only three days prior to its signing, allowing no input to challengers.

33. *O Estado de São Paulo*, Aug. 23, 1979.

34. Joint ventures imply sharing of costs, and of technical, marketing, production, and managerial skills. They may include free licensing of technology from one partner to another, the supply of packaged technology while withholding control, and other mechanisms.

35. Syllus (1982).

36. Nuclear trade has enjoyed a special status under GATT and other multilateral trade arrangements, escaping restrictions on governmental intervention. Thus, preferential treatment of local suppliers has been allowed, as have subsidies, through R&D funding, acceptance of higher prices for domestically supplied equipment, and fiscal measures (Walker and Lonnroth 1983).

37. The diffusion of knowledge or information in nuclear science, as opposed to a particular technology, cannot be contained.

38. Potter (1982); Scheinman (1987).

39. Quester (1981). The NPT establishes a bargain by which signatories agree not to transfer nuclear weapons to nonnuclear weapon states, not to produce them (unless they had them already), and not to export nuclear materials without safeguards. In turn, the NPT acknowledges—under Article IV—the right of all parties to develop nuclear energy for peaceful purposes and to participate in the exchange of nuclear knowledge, equipment, and materials associated with peaceful uses (R. Smith 1987).

40. Such safeguards point to a full coverage of a country's present and future nuclear facilities by IAEA inspectors. The commitment to submit all of one's facilities to international inspections is undertaken through ratification of the NPT, or the Tlatelolco Treaty (de jure) and similar regional counterparts, or through a variety of trilateral agreements.

41. Officially known as the Nuclear Suppliers Group, the so-called London Club was organized in 1974–75 (Nye 1981; Timerbaev and Watt 1995).

Canada, West Germany, France, Japan, the United Kingdom, United States, and Soviet Union were the seven original participants. Eight other countries declared their allegiance to the norms in 1978, and with subsequent additions, the group grew to 30 states by 1994.

42. Other newly industrializing countries that had embarked on nuclear power programs by the 1970s include Taiwan, South Korea, Mexico, and India.

43. Apart from these multilateral restrictions, West Germany unilaterally interdicted any further export of reprocessing plants in 1979.

44. South Africa and South Korea, among others, were similarly successful in obtaining sensitive technologies from established suppliers after 1975.

45. Heavy water was included in the list of sensitive technologies approved by the London Suppliers Group only in 1976. On pressures in Brazil for heavy water, rather than light water, reactors, see Chapter 5.

46. Skidmore (1988). In 1954 the United States had prevented Brazil's acquisition of ultracentrifuge (enrichment) equipment from West Germany and forced the ousting, a year later, of Admiral Alvaro Alberto, an advocate of an independent nuclear program.

47. These transfers were to be accompanied by standard trilateral (non-full-scope) safeguard agreements with the IAEA.

48. For an evaluation of the success of technology control in the 1970s, see Walker and Lonnroth (1983).

49. The main difference between a nuclear plant and a conventional steam plant is the nuclear steam supply system, where a nuclear reactor and other components replace the boiler (Surrey and Walker 1981).

50. Encarnation (1989:20) defines bargaining power as the ability to improve the range of plausible outcomes and to improve the probability of securing the preferred outcome. On bargaining power, see also Caporaso (1978).

51. Kobrin (1987); Vernon (1977:171).

52. Newfarmer (1985).

53. This brief review of the international nuclear market draws on Walker and Lonnroth (1983).

54. The structure of the international heavy electrical machinery market (particularly power plant equipment) is analyzed in Newfarmer (1985) and Surrey and Walker (1981). The major international producers included Siemens (FRG), CGE (France), ASEA (Sweden), Brown Boveri (Switzerland), G.E. Co. (U.K.), G.E., Westinghouse, and ITT (U.S.), and Hitachi, Toshiba, and Mitsubishi (Japan).

55. Surrey and Walker (1981). Although some industrializing states regard the London Suppliers Group as a cartel, this so-called club plays less of a traditional economic role (such as price-fixing or market sharing) and relies on an intergovernmental political understanding to strengthen the nonproliferation regime.

56. U.S. firms argued that the government's nonproliferation efforts had contributed, to some extent, to the erosion of their own prior monopoly.

57. Wonder (1977).

58. Ibid., p. 284.

59. This was the case, for instance, with the Canadian-Romanian agreements.

60. KWU was born out of a joint venture between Siemens and AEG, two of the four largest German producers of heavy electrical equipment.

61. *Relatório* 6, no. 5, pp. 17–18.

62. Grabendorff (1981).

63. On adaptation by MNCs, see Schatz (1987).

64. On how the particular arrangements of national financial systems limit both the marketplace options of firms and the administrative choices of government, see Zysman (1983).

65. In Brazil, major state enterprises such as Petrobrás, Eletrobrás, Siderbrás, and CVRD relied on private foreign banks for 17 percent of their investment in the early 1970s, with internal resources accounting for between 30 and 50 percent. Levels of self-financing went down to about 25 percent by 1980. Firms in the electrical utility sector absorbed vast shares of public and foreign resources, the latter representing about 30 percent from their total borrowing (Trebat 1983:88, 207, 211).

66. *Jornal do Brasil*, Dec. 19, 1976, p. 42.

67. *O Globo*, July 15, 1976, p. 30. Bankers from the Dresdner Bank, Deutsche Bank, Commerzbank, and Kreditanstalt accompanied representatives of KWU and associated firms at the signing of the nuclear agreement in June 1975 (*Manchete*, July 12, 1975, pp. 10–12). German suppliers' bids were backed by their country's export credit agencies, offering up to 60 percent of the project's cost, under exceptionally generous terms. A decline in budgetary assistance to export financing among European suppliers made the financing of capital goods exports more disciplined and restricted by the mid-1980s than they had been ten years earlier when the agreements were signed. For an analysis of financing of nuclear power programs, see Mounet (1985), and Charpentier and Bennett (1985). The World Bank excludes nuclear power from the energy projects it funds.

68. *Jornal do Brasil*, Dec. 19, 1976, p. 42.

69. On the initial reluctance by state financial agencies (Central Bank, Banco do Brasil, and the Ministry of Finance) to disburse funds for the alcohol program, see Barzelay (1986).

70. On how foreign financing strengthened the local private industrial bourgeoisie, see Becker et al. (1987). On state entrepreneurship and external borrowing in Brazil, see Frieden (1991).

71. Sábato (1979); Tanis (1985).

72. Grant-back clauses imply control over innovations. Improvements and R&D results are returned to the supplier.

73. Tie-in sales are clauses according to which the recipient has to obtain materials and spare parts from the vendor.

74. Tie-out clauses forbid the acquisition of complementary or competing technologies from any other supplier (Frame 1983).

75. Clause 14 in the shareholders' accord (*Jornal do Brasil*, Aug. 26, 1980).

76. *Relatório* 3, p. 113. On the high incidence of tie-in imports in Brazil's complex capital goods sector, see Chudnovsky and Nagao (1983:128).

77. *Relatório* 3, p. 124.

78. *Relatório* 2, p. 126, and *Relatório* 4, p. 73. By 1995 the technology supplied by KWU would be considered obsolete. In reality, by 1994 not a single plant (of the eight planned) was operating.

79. *Relatório* 3, p. 88; *Jornal do Brasil*, Aug. 26, 1980. According to Chudnovsky and Nagao (1983:128), formal restrictions on exports were found in 17 percent of all agreements on complex capital goods.

80. *Relatório* 3, pp. 91–92.

81. Walker and Lonnroth (1983).

82. Relatório 3, p. 112.

83. *Relatório* 3, p. 108; Carvalho (1981).

84. The jet-nozzle process consumes 15 times more energy than gaseous diffusion and 45 times more energy than ultracentrifuge techniques, both of which are alternative enrichment technologies.

85. This effect is discussed in Frame (1983).

86. These advantages that joint ventures offer to suppliers are discussed in Charpentier and Bennett (1985).

87. On foreign investors' preferences, see Courtney and Leipziger (1974). On joint ventures, see also Frame (1983).

88. Sagafi-Nejad et al. (1981).

89. In a comprehensive study of Brazil's private sector, Villela and Baer (1980) encountered similar levels of dissatisfaction with foreign partners reluctant to transfer technology but eager to exercise total control of the project. On India's relative aversity to joint ventures, see Encarnation (1989).

90. Syllus (1982).

91. Sklar (1976).

92. Dahlman et al. (1985:ii). This section's considerations about the relative effectiveness of technology transfer mechanisms borrows from the same source.

93. See, for instance, Gereffi (1983).

94. *O Estado de São Paulo*, July 21, 1985, p. 38, and May 4, 1986.

95. The fuel element fabrication plant and the conversion plant accounted for $75 million; the uranium concentrate plant for $260 million; mineral exploration for $150 million; and the enrichment pilot plant for $300 million. The initial cost of NUCLEP, the joint venture responsible for producing sets of heavy components for nuclear plants, was $300 million. (*O Estado de São Paulo*, July 17, 1985, p. 22).

96. *Relatório* 4, p. 220, and *Relatório* 5, p. 71; *Brasil Energy*, July 10, 1980, p. 4.

97. *O Estado de São Paulo*, Nov. 20, 1990, p. 16. At least $2.9 billion had been invested in the first two plants under the agreement—Angra 2 and Angra 3—by 1984 (*O Estado de São Paulo*, Oct. 29, 1986, p. 31; *Jornal do Brasil*,

May 15, 1985; *Veja*, Dec. 12, 1984, p. 102; *O Globo*, June 27, 1985, p. 23).
About $5 billion had been spent on Angra 2 alone by 1994, and an additional
$2.3 billion was necessary to complete it (*JPRS*, April 14, 1994, p. 11). Nucle-
brás officials argued that the first plants in the series would reflect higher
costs, whereas succeeding plants would reap the benefits of cumulative
learning and production processes.

98. *Relatório 4*, p. 81.

99. *Relatório 5*, pp. 27–29, and *Relatório 4*, p. 89. In the committee's report, overpricing is attributed specifically to Siemens's and KWU's membership in the International Electrical Association (*Relatório 6*, no. 6, pp. 14 and
57, and *Relatório 3*, p. 178).

100. Newfarmer (1985:126).

101. *La Razón*, Feb. 21, 1985, p. 20, and May 26, 1985, p. 6; *Nuclear Engineering International*, Nov. 1985, p. 7; *Energeia*, Aug. 1984, p. 39; Poneman
(1987:180).

102. On Argentina's energy sector, see Guadagni (1985). On Brazil's nuclear program in the context of its energy balance, see Erickson (1981b) and
Solingen (1991).

103. *Relatório 4*, p. 83. The capital cost of Argentina's CANDU reactor
was comparable to light water reactors, but the fuel cycle related to CANDU
reactors was projected to be half as expensive as their light water counterparts (*Realidad Energética*, Apr. 1983).

104. On how importing embodied technology (such as capital goods) is assumed to transfer the multiplier effects of investments—the Schumpeterian
"windmill profits"—back to the supplier, see Evans (1979). For similar arguments, see Cooper (1972), Herrera (1972), Sunkel (1973), Furtado (1976),
and Mytelka (1979).

105. These costs included a few domestic contracts. See *Jornal do Brasil*,
May 15, 1985; *O Globo*, June 27, 1985. Capital goods include the reactor and
turbines, and mechanical, electrical, and control equipment.

106. *Relatório 4*, p. 176. Engineering costs often include: (1) project management (planning, basic technical criteria); (2) viability studies and siting
(geology, hydrology, etc.); (3) design and engineering (civil, mechanical, electrical); (4) architecture; (5) consulting and special studies; (6) quality control.

107. *O Estado de São Paulo*, July 17, 1985, p. 22.

108. *El Desarrollo Nuclear Argentino*, p. 196.

109. Both are rather optimistic readings of technology costs. Erber (1981a)
notes that direct costs of technology imports are merely the "tip of the ice-
berg," and tend to overlook less visible costs, often hidden in payments for
training, plans, designs, foreign personnel, and manuals.

110. CNEA devoted an average of 10 percent of its total budget for R&D
(*Energeia* 47/8, 1984, pp. 17–19; CNEA, *Memoria Anual*, 1981, 1982, 1983).
Its budget proposal for 1985 suggests that payments to KWU amounted to
less than 1 percent of the budget (*Energeia* 52, 1985, p. 17). CNEA's annual
budget averaged $1.5 billion to $1.8 billion between 1975 and 1983. CNEA's

general expenditures were beyond public scrutiny, and the program was at least partially funded in ways not directly reflected in the federal budget (*Nucleonics Week*, May 28, 1987).

111. *O Estado de São Paulo*, June 7, 1985.

112. *FBIS*, Jan. 4, 1991, p. 15.

113. *Nucleonics Week*, June 2, 1988, p. 9. It represented $1.5 billion in 1982, or 3.4 percent of the $44 billion foreign debt at the time (*Clarín*, Sept. 30, 1984, p. 6; *Nucleonics Week*, Dec. 1, 1983). According to another estimate, the nuclear program accounted for $5 billion of the $70 billion public foreign debt (*Financial Times*, Nov. 25, 1985). Yet another estimated CNEA's external debt by 1987 to be $10 billion (*La Prensa*, July 12, 1987).

114. *Relatório* 5, p. 36. NUCLEP's technical committee, for instance, was to be transferred to Nuclebrás as majority partner in December 1986 (*Relatório* 3, p. 126). Senator Itamar Franco and three other opposition senators resigned in 1980, owing to the government's opposition to questioning General Cavalcanti on controversial secret documents related to the nuclear agreement.

115. *Senado Federal*, Projeto de Decreto Legislativo No. 3, 1983 (800/4/83). Centro Gráfico do Senado Federal, Brasília. See also *O Estado de São Paulo*, Sept. 8, 1985, p. 10.

116. On the "obsolescing bargain" and the declining advantages of multinationals with the project's life cycle, see Vernon (1971) and Grieco (1984). On the alternative perspective, see Gereffi (1983) and Bennett and Sharpe (1979).

117. Gerschenkron (1962).

118. On how intense rivalry among suppliers allows recipients high levels of national content, see Grieco (1984), Gereffi (1983), Encarnation and Wells (1985), and Doner (1992). High concentration and fierce competition for exports was also characteristic of the automobile industry in the 1950s and 1960s (Bennett and Sharpe 1979).

119. Amsden (1992).

120. For a similar emphasis on domestic structural and institutional constraints in bargaining processes, see Bennett and Sharpe (1979), Grieco (1984), and Kudrle (1985).

121. On how low uncertainty about the investment strengthens the recipient, see Moran (1978, 1985).

122. On the importance of technical capacity to screen investments, see Stepan (1978), Moran (1978, 1985), and Packenham (1992).

123. On "structural power," see Caporaso (1978). I use it here to point out that levels of consensus and autonomy underwrite the rules shaping the bargaining context: who has the right to bargain, over what scope of issues (local content, pricing, technology transfer), and with what degree of independence.

124. According to Newfarmer (1985:126), developing-country reliance on single suppliers of heavy equipment resulted in overpricing at nearly 50 percent above actual prices.

125. The point is emphasized in Moran (1978) and in Bennett and Sharpe (1979).

126. On the financial power of multinational corporations as "pivotal bargaining power," see Klapp (1987).

127. The former Federal Republic of Germany, France, Great Britain, and Japan adopted the basic U.S. design of pressurized water reactors.

128. On the analytical weakness of rhetorical categories such as "autonomous," "self-reliant," and "dependent" policies, and the utility of studying actions, processes, and outcomes, see Katzenstein (1978) and Tooze (1984).

Chapter Five

1. Jair Mello, from the thorium group (*Relatório 6*, no. 5, p. 361).

2. Physicists have played important roles in the development of nuclear industries in industrializing countries. Most technological research in Brazil has been in the hands of physicists (*Boletím Sociedade Brasileira de Física* 1, no. 9, 1978, p. 11). Moreover, physicists arguably led the most vocal institutional public critique of the nuclear program. This chapter also refers to nuclear engineers, such as Joaquim Carvalho and Jair Mello, who formed part of the core group of activist scientists. There can be a significant degree of overlapping in the endeavors of nuclear physicists and engineers. This chapter builds on extensive interviews conducted in April and May 1985 with leading members of Brazil's scientific community.

3. A scientific community can be defined as an "organized group with a developed system of beliefs, with a developed system of institutions for internal communication, as well as a system of communication for dealing with other social groups, and which is bound by certain traditional norms of behavior for furthering their individual and collective work in science" (Dedijer 1968:162).

4. *Manchete*, Feb. 26, 1977. This assessment was informally confirmed by personal interviews with "neutral" members of the international physics community. Systematic research on nuclear physics started in Brazil in 1934, in the Faculty of Philosophy, Science, and Letters at the University of São Paulo, where two Italian physicists, Gleb Wataghin and Giuseppe Occhialini, trained the first group of Brazilian physicists (Rowe 1969). For a comprehensive study of the social history of Brazil's scientific community, see Schwartzman (1991).

5. Souza Santos and Pompeia, in collaboration with Wataghin, discovered the showers of penetrating particles in cosmic radiation (1940). Mário Schemberg collaborated with George Gamow in the formulation of the theory of the neutrino collapse of stars (1940) (Lopes 1984).

6. Goldemberg (1978); Biasi (1979).

7. Particle, nuclear, and solid-state physics were driving forces in the development of high technologies such as aerospace, electronics, telecommunications, and nuclear fusion.

8. Levy-Leblond (1976:136).

9. The First Basic Plan for Scientific and Technological Development (1972–74) allocated generously to the development of Brazilian science and technology.

10. Personal interviews, Rio de Janeiro, April–May, 1985.

11. Personal interview, Rio de Janeiro, Apr. 10, 1985. See also *Tendencia*, Mar. 1984, pp. 40–42.

12. See inquiry by a parliamentary committee on the nuclear program (*Relatório* 3, p. 103).

13. Physicist Marcelo Damy was Director of CNEN in those years. Brazil's leading research programs pursued these technical options at the São Paulo Instituto de Energia Atômica and the Instituto de Pesquisas Radioativas in Belo Horizonte.

14. Although most first-rate Brazilian physicists did not participate in the nuclear program, it would be inaccurate to claim that all those who did, or who supported it, were mediocre scientists, as was sometimes implied (see, for instance, *O Estado de São Paulo*, Apr. 2, 1978). General Dirceu Coutinho, former director of NUCLEI (Nuclebrás subsidiary for uranium enrichment) was highly respected professionally among his peers (*Relatório* 6, no. 6, p. 64.). Hervásio de Carvalho was considered a pioneer physicist, even by those who criticized his role as director of CNEN (Lopes 1972:60). The Parliamentary Investigating Commission cites Carvalho as the first Ph.D. in nuclear engineering worldwide (*Relatório* 6, no. 4, p. 261). His successor, Rex N. Alves, was a graduate of the Sorbonne and author of numerous professional articles.

15. See Lepecki (1985), who was a manager at NUCLEN. In December 1974, the Division of Technology and Development at CBTN (soon to become Nuclebrás) had a total of six hundred scientists and engineers (*Correio Brasiliense*, Mar. 24, 1975, pp. 4–5). During its first years Nuclebrás had about 45 nuclear physicists (nine Ph.D.'s and four M.Sc.'s) (*Boletím S.B.F.*, 1975, p. 9; *Veja*, Mar. 8, 1978). In 1984 the total number of engineers and technicians at NUCLEN (Nuclebrás Engineering) was 445.

16. *Clarín*, Sept. 30, 1984.

17. *Relatório* 6, no. 3, p. 412. See also deposition by Professor Roberto Hukai, *Relatório* 6, no. 5, p. 133.

18. A Brazilian physicist estimated that number at ten (high- to medium-level professionals), including Hervásio de Carvalho (director of the Atomic Energy Commission), Ricardo Brant Pinheiro, Trentino Polga, Amaranti, and Ivo Jordan (personal interview, May 1985). On the exclusion of most scientists, see deposition by David Simon, *Relatório* 6, no. 6, p. 79. Price (1965: 83) defined as "insiders" those eminent scientists who hold important governmental positions and "are likely to accept subordination of science to the value systems established by the nation's political traditions and interpreted by the authority of its government." An interesting pattern of flows from the "inside" out took place. David Simon participated in consultations leading to the Protocol of Brasília (1974), in initial negotiations between the utility

Furnas and the German supplier KWU, and was invited (but refused) to assume a directorship at NUCLEN, leaving Furnas in 1979. Sérgio Salvo Brito, who assumed CNEN's directorship, left after a very brief service. Directors of NUCLEN (Nuclebrás Engineering) and NUCLEI (Nuclebrás Isotopic Enrichment), Joaquim Carvalho and General Dirceu Coutinho, left in 1979 following repeated attempts to introduce changes in the program.

19. Price (1965:84) defined "outsiders" as the independent critics of public policy, "less willing to accept the validity of the traditional political ethos."

20. On how vocal scientists tend to express opinions backed by a broader slice of their community, see Gilpin (1962).

21. In the midst of World War II (1942) the Brazilian Navy, having lost several ships in encounters with German and Italian submarines, called upon a group of physicists at the University of São Paulo (including Marcelo Damy) to develop sonars for the Navy's use. Following a successful switch from purely scientific to technological research, the Navy discontinued its interest in this type of technological research at the end of the war, and the physicists went back to their academic activities (Schwartzman 1978:258).

22. José Leite Lopes and Mário Schemberg were prominent political activists. Schemberg had been elected by the Communist Party to the Constitutional Assembly of 1946 and lost his mandate when the party became illegal in 1947.

23. Rowe (1969:106). These included Italian and German scientists such as Wataghin, Occhialini, and Gross.

24. Most details of the nuclear agreements with West Germany were kept from the scientific community. In a secret document circulated among the internal security services and later leaked to the press, the following scientists were singled out as the main opposition to the nuclear program: Mário Schemberg, José Goldemberg, Luiz P. Rosa, Rogerio Cerqueira Leite, Jair Mello, Marcelo Damy, Vitor Cleo, Luiz C. Menezes, Ennio Candotti, José Leite Lopes, Alfredo Taculine, Hernani L. Amorim, and a few others (*Folha de São Paulo*, June 6, 1980). See also Nussenzveig (1969). In accordance with Complementary Act No. 75, anyone considered affected by the Institutional Acts (which limited political activity) was also to be discharged from any governmental, teaching, or research institution (Sant'anna 1978; Ames 1987).

25. Restrictions known as *cassações brancas* applied to a large number of scientists and included limits on participation in international scientific meetings, conducting research abroad, conveying ideological statements, and obtaining funding and fellowships. The regime imprisoned scientists with any real or perceived ties to the political left (*Boletím S.B.F.* 3, 1975). Professor Paulo Miranda, from the University of Brasília, was dismissed for obtaining his B.Sc. and M.Sc. in the USSR, which Brazil did not validate at the time (ibid., 3, 1978).

26. Polanyi's Republic of Science assumes a universal character to the sci-

entific enterprise, an emphasis on the objectivity of science and its quest for truth and the rejection of external control in topic selection. Many among the Brazilian scientists interviewed rejected the notion of science as truthful, rational, objective, neutral, and "for its own sake," which was more prevalent among their North American counterparts (Ziman 1982).

27. Among them were Luiz Pinguelli Rosa, José Goldemberg, Jair Mello, J. Moreira, and Mário Schemberg. Schemberg, former president of the Brazilian Physics Society, criticized the government for concentrating on "monumental energy projects" that resulted only in an increase in the country's foreign debt, citing the hydroelectric and nuclear programs as examples (*Brasil Energia*, Nov. 10, 1981, p. 6).

28. Published in *O Estado de São Paulo* and quoted in Soares (1976).

29. The difference between basic and applied research is blurred in the actual practice of science. Basic science results in new ideas and can affect economic growth.

30. *Boletím S.B.F.*, 1978.

31. Street (1981).

32. Goldemberg (1978:11).

33. Previously, dismantled research teams working on alternative nuclear paths (heavy water, thorium, fast breeders, and ultracentrifuge techniques) went back to work (*Brasil Energy*, Mar. 25, 1982, p. 6). Neglected university institutions received full support after years of operating with scarce resources, particularly compared to those devoted to the nuclear industry. Former critics, like José Goldemberg, praised the creation of a National Physics Center for Plasma and Controlled Fusion Research as a positive step in the development of a national program.

34. *Folha de São Paulo*, Oct. 25, 1983, p. 8.

35. José Goldemberg became Science Minister.

36. For the most important critiques of the program by scientists, see Rosa (1978, 1984), D. Simon et al. (1981), and Goldemberg (1978). Additional critiques are included in Mirow (1979) and Girotti (1984). There was additional opposition to the program on other grounds, such as the underestimation of hydroelectric resources, and environmental and safety issues.

37. Leading physicists, such as José Leite Lopes, José Goldemberg, Marcelo Damy, and others, advocated the thorium option (*O Estado de São Paulo*, Apr. 25, 1976). The thorium (Th-U-233) fuel cycle had some technical drawbacks but also had the advantage of furthering local industrial development and domestic resources utilization, in light of the technology's alleged relative simplicity (Brito 1967:216).

38. Morel (1979:118). The thorium group included nuclear engineers, physicists, mathematicians, and chemists. The total number of high- and medium-level personnel at the Institute's various departments in 1969 was 220. (*Relatório* 6, no. 4, p. 423).

39. Emphasis on fuel independence was the "spinal cord," in scientist Jair Mello's terms, guiding the technical conceptualization of this reactor (*Relatório* 6, no. 5, p. 362).

40. An array of scientific publications and conference papers came out of this project. A book edited by Mello, Urban, Brito, and Lages published in 1966 (*Introduction to Nucleoelectric Generation*) attracted the interest of a major international publisher (*Relatório* 6, no. 5, p. 354). In addition, the results of research at Siemens on thorium reactors (submitted in a 1967 IAEA conference) closely resembled those obtained by the Brazilian team, thus confirming the reliability of its research (personal interview, May 1985, Rio de Janeiro). It should be noted, however, that most advanced countries had not found thorium cycles a commercially viable technology.

41. Only heavy water and graphite were imported, from the U.S. and France, respectively (*Relatório* 6, no. 4, p. 402).

42. Physicist Hervásio de Carvalho was central to that alignment and unconcerned with the reliability of the U.S. supply of enriched uranium (*Relatório* 6, no. 5, p. 372).

43. D. Simon et al. (1981:66); *O Estado de São Paulo*, Oct. 19, 1983, p. 6. Others argued that this program might have required up to 25 years (*Relatório* 6, no. 1, p. 311, and *Relatório* 6, no. 5, p. 157).

44. Dickson (1978:578). Physicists could, for instance, play important roles in safety analysis, radiation control, nuclear instrumentation, quality control, treatment of radioactive waste, reactor physics, and materials analysis (*Boletím S.B.F.* 5, no. 6, 1975).

45. For a similar perception of what may have been an apparent (rather than real) staff shortage in the U.S. nuclear energy program, see Kuhn (1966:41). Many Brazilian scientists agreed that there was no critical mass for the *fast and immediate* development of an indigenous program.

46. About 31 Ph.D.'s and 15 M.Sc.'s were working in nuclear research projects at academic institutes and within Nuclebrás itself (*Boletím S.B.F.* 5, no. 6, 1975). A later report adjusts the initial figure of 277 to 366 Ph.D.'s in 1975, in addition to 184 doctoral candidates and 141 M.Sc.'s (*Boletím S.B.F.* 1, no. 9, 1978). The Second Basic Plan for Scientific and Technological Development published in 1976 confirmed 330 Ph.D.'s and 260 M.Sc.'s. (*II Plano Básico de Desenvolvimento Científico e Tecnológico*. São Paulo: Sugestões Literarias, 1976, p. 153).

47. *II Plano Básico*. The III Basic Plan (1980–85) reports seven hundred Ph.D.'s in physics in Brazil by 1984.

48. *Boletím S.B.F.*, 1975, p. 9. In fact, many French physicists were replaced by engineers and technocrats in the early 1950s (Jasper 1990:74).

49. Dickson (1978:579).

50. Nordlinger (1981) discusses these tactics in a more general context.

51. Falicov (1970:41).

52. The first generation of Argentine physicists was trained, as in the case of Brazil, by German refugees who fled the Nazi regime (Falicov 1970). The founding father of the Bariloche center, José A. Balseiro, was a disciple of exiled Austrian physicist Guido Beck, who had also trained Brazilian physicists.

53. Despite at least eight "disappeared" in the 1970s, CNEA (and the

Navy) had arguably been able to protect its scientists from the worst waves of military repression. Scientists involved in Army-controlled projects were far less sheltered.

54. Tellez (1966); Westerkampf (1982).

55. Street (1981). About 40 percent of the Physics Society membership was estimated to have lost their jobs, or to have fled the country, or both. The Institute of Geo-Helio Physics, along with other theoretical research centers, was dissolved (Levín 1981).

56. According to a Brazilian physicist, scientists were more tolerated in Brazil because they were perceived to have little contact with grass-roots political movements (personal interview, São Paulo, May 25, 1985).

57. Westerkampf (1982).

58. Lopes (1972, 1978); Cooper (1974); N. Stepan (1976); Sant'anna (1978); Sagasti (1979). For a neoclassical critique of the emphasis on developing national science, see Nelson (1974).

59. However, Brazil's heavy funding of science and technology in the 1970s questions the claim that greater economic integration with the outside world always weakens support for domestic science.

60. According to Morel (1979:38), Alberto was dismissed for his nationalist initiatives in nuclear policy.

61. Nussenzveig (1969).

62. Street (1981:22).

63. Shils (1972).

64. Lakoff (1977:373). Experimental scientists are regarded as professionally the most rigorous, having a greater degree of consensus over the goals of research and over what is worthy of investigation (Ben-David 1972).

65. Barber (1952); Mulkay (1977).

66. To some extent, various studies of student activism in Brazil and Argentina validate these claims. Activism from the left gained relatively greater support in letters and sciences (including physics) than in engineering, medicine, pharmacy, and other technically oriented departments. Ladd and Lipset (1972) have discerned similar patterns elsewhere.

67. Theoretical physics has been traditionally more developed in Brazil than experimental physics, as in most places where expensive experimental equipment is out of bounds for resource-poor academic institutions.

68. For an overview, see Lakoff (1977).

69. Nelson (1974). For a pioneering study, see J. Lewis and Litai (1988).

70. For a more systematic effort to understand the variability of state-scientists relations across political and economic systems, developed and underdeveloped, see Solingen (1994a).

71. For some parallels with the French and Japanese programs, see Jasper (1990), and Baumgartner and Wilsford (1994), and Low (1994), respectively. On the limited influence of French scientists within the state, see Gilpin (1968:371). Japan and France enjoyed similarly high levels of macropolitical consensus over industrial policy at the time when their nuclear industries

were designed. MITI imposed that consensus over a segmented sectoral decision making in Japan (Low 1994:21–22).

72. J. E. Katz (1982).

73. On this "symbiosis of power" between scientists and the state, see Clark (1987).

74. Haberer (1969:303).

75. Polanyi (1968); Merton (1973).

76. On the role of external determinants of scientific activity, see Merton (1968).

77. On the centrality of physicists in important political debates across political and economic systems, see Solingen (1994a).

78. Pinto (n.d.), for instance, argued the physical sciences are analytic, deductive-nomological, and based on philosophical postures regarding the understanding of the world. Chemical and biological sciences, instead, are more geared to the identification of structures and processes, disconnected from major philosophical interpretations. These traits are arguably related to political proclivities.

79. Chemists and biologists can be more easily absorbed by private industry.

80. Physicists were far more politicized than mathematicians, biochemists, and others, who, according to some Brazilian physicists (personal interviews, Rio de Janeiro, April 1985), were more reluctant to abandon the ivory tower. Proportionally, there are far more chemists than physicists in Brazil, but they have been much less politically active, according to a biologist (personal interview, São Paulo, May 25, 1985). However, governmental harassment of scientists did not exclude biologists and chemists from the Instituto Oswaldo Cruz and other academic centers.

81. The prominence of physicists (relative to other disciplines) among the Argentine "disappeared" and persecuted can be deduced from the lists produced by Amnesty International and by the American Association for the Advancement of Science. Among these victims, former CNEA physicist and director of the National Institute of Industrial Technology, Máximo Victoria, argued that physicists were singled out because of their outspoken demands, during the years preceding the 1976 coup, for a reorientation of the country's science and technology policies toward a more independent path (AAAS Committee on Scientific Freedom and Responsibility. *Clearinghouse Report on Science and Human Rights*, 1981 and 1984).

82. Meneses and Simon (1981:3). Brazil's Physics Association approved a charter warning all physicists to reject employment in any program geared to the production of nuclear weapons (published in *O Estado de São Paulo*, Dec. 10, 1983, p. 5). However, even leading critics of both nuclear programs were not opposed to the achievement of a national enrichment and reprocessing capability or to the development of thorium and fast-breeder reactors, despite the intrinsic proliferation potential of these technologies (*Energia Nuclear*, 1978, pp. 13–17; *O Estado de São Paulo*, Apr. 25, 1976).

83. Some have even advanced the right to conduct peaceful nuclear explosions. As can be recalled from Chapter 4, the London Suppliers Group approved norms and regulations limiting the diffusion of sensitive technologies (like uranium enrichment and reprocessing) to additional countries.

84. For a fascinating account of the leadership of the People's Republic of China's ambiguity toward scientists, see Lewis and Litai (1988).

85. Arguably, there were elements within the Security Services who believed that scientists should be tolerated for strategic reasons (interview with Brazilian scientist, São Paulo, May 24, 1985). It was also reported that while the Air Force was more combative in its position vis-à-vis the scientists, the Army agreed, to some extent, with their criticism of the nuclear program (Stone 1977:3).

86. Rosenberg (1982:13); Clark (1987).

Chapter Six

1. Lall (1991).

2. For a comprehensive study of Brazil's public sector, see Trebat (1983). On state entrepreneurship and technological development more generally, see Marton (1986).

3. Cortés and Bocock (1984).

4. Syllus (1982).

5. This would imply a parallel development of domestic suppliers (Sábato et al. 1978; Tanis 1985).

6. At this stage the local partner has to assume 50 percent of the technological risk.

7. Innovations can be defined as "the first 'commercial introduction' of a new product or process" (Clark 1985:71).

8. *Relatório* 3, p. 132.

9. Mechanical systems are defined as "a group of components and pipes conveying a working medium which form a well-defined functional unit" (Lepecki 1985:11).

10. *Noticias Nucleares* (KWU), no. 9, p. 5.

11. A reactor system encompasses the nuclear and thermohydraulic design of the reactor core (Lepecki 1985:12).

12. These areas included the reactor coolant system, reactor instrumentation and control, core design, and safety analysis (Lepecki 1985).

13. Vargas (1976). Many interviewees, including those employed by Nuclebrás, concurred in their private revelations that German officials were very reticent in transferring technology.

14. Basic engineering implies knowledge of all functional requirements, preliminary calculations, flow diagrams, fundamental requirements and limitations in operation, complete design of processes, thermal balances, location of components and equipment, and materials specifications. Detailed engineering includes final calculations and plans, selection of materials,

composition of specifications chart for the acquisition, fabrication, erection, and inspection of systems, structures, and equipment, and technical evaluation of offers (Sábato et al. 1978).

15. *O Estado de São Paulo,* July 6, 1985.

16. During his tenure at NUCLEN's Department of Industrial Promotion, Carvalho attempted to coordinate the participation of IPT and IPEN with NUCLEN and the private sector. He favored sending a group of specialists in material science and welding technologies to Germany to learn from German industrial firms and from governmental services to industry. KWU representatives within NUCLEN, apparently lobbying other Brazilian directors into rejecting Carvalho's initiative, vetoed such efforts (Carvalho 1981a, 1985).

17. Lepecki (1985). NUCLEN had 821 employees in 1981, including 350 engineers and technicians, 130 of whom were trained in West Germany by KWU (*Relatório* 4, p. 202; Backhaus and Wildgruber 1982).

18. Such conditions include seismic and environmental criteria, cooling water systems, external influences, frequency of electricity grids, and voltage levels (Fabricio 1982).

19. Lepecki (1985).

20. Rosa et al. (1984). Even a staunch critic of the Brazilian-German accord, physicist Pinguelli Rosa, suggested that a restructured and nationalized NUCLEN could supply expertise and equipment to the energy sector and could help local firms perform in areas that generally demand foreign expertise (Rosa 1985).

21. Despite repeated formal requests to obtain information on technological development at NUCLEP in 1985, Nuclebrás's public relations office at the time refused to grant permission either to interview NUCLEP's officials or to visit their facilities.

22. NUCLEP's annual capacity was to be one set of heavy components for a pressurized (light water) reactor, beginning with the third plant (Iguape). Its personnel of 382 (1978) grew to over 1,300 in 1982 (98 percent Brazilians), of which 1 percent was trained in Germany (*Relatório* 4, p. 209). German participation in NUCLEP shrank to less than 3 percent by 1985 (*Jornal do Brasil,* May 15, 1985) before NUCLEP's dissolution.

23. Nuclebrás retained the remaining 90 percent (Nuclebrás, Secretary of Public Relations, 1982). NUCLEMON (Nuclebrás de Monazita e Asociados Ltda., 100 percent Nuclebrás) was responsible for prospecting, extraction, research, and production of heavy radioactive sands (such as monazite, ilmenite, and rutile, and subproducts such as uranium, thorium, and zirconium) and rare earths.

24. Known uranium reserves grew from close to 11,040 tons in 1974 to 301,490 tons ten years later, an amount suitable for supplying 48 plants of 1,300 MW for 30 years. The energy potential of these reserves was equivalent to 35 times the reserves in oil, gas, coal, and charcoal combined (Nuclebrás, *O avanço,* n.d.).

25. The French firm Pechiney Ugine-Kuhlman supplied basic engineering, and Brazilian firms Natron and Andrade Gutierrez were responsible for detailed engineering and civil construction, respectively. Domestic firms also supplied almost 90 percent of the equipment. The rest were special equipment and instrumentation not yet produced in Brazil (Nuclebrás, *O CIPC,* n.d.). Participation in the Poços Complex improved one firm's basic design capability, allowing it to export a similar project and to conduct the feasibility study and definitive estimate for the construction of a new mining and yellowcake production complex in Lagoa Real.

26. RBU demanded to import a portion of the equipment, but Brazilian services (construction and assembly) and equipment accounted for 90 percent of total investments (Nuclebrás 1982a).

27. Inputs were imported enriched uranium from Germany and zircalloy tubes from Argentina. At a later stage, local pellet fabrication and UO_2 reconversion were to follow (Lepecki 1985). About 200 people were employed at this plant in 1982, 52 of them technicians trained in Germany (at RBU's plant) for a period of two years.

28. The architect-engineers of this facility were from Steag and KWU's subsidiary Interatom. Brazil's participation in the supply of equipment and machinery for NUCLEI was to reach 40 percent (*O Globo,* Sept. 7, 1978). NUCLEI grew from 92 engineers and technicians in 1978 to 248 in 1982 (9 percent German). During these years an average of 6 percent of its Brazilian personnel was trained in Germany (*Relatório* 4, p. 193).

29. KEWA/UHDE were to provide basic and detailed engineering, civil construction supervision, assembly, and start-up testing. By 1983 the project was still in the design phase (*O Estado de São Paulo,* Dec. 10, 1983, p. 5).

30. The research institute IPEN reportedly operated a reactor for production of fissile isotopes (such as plutonium) and conducted research in uranium enrichment through lasers and ultracentrifuge techniques (*O Estado de São Paulo,* Oct. 4, 1983, p. 10, and Oct. 12, 1984, p. 3; *O Globo,* Dec. 3, 1983, p. 18; *Folha de São Paulo,* May 11, 1985, p. 24). Research on reprocessing, fusion, laser enrichment, and "breeder" reactors allegedly took place at the Aerospace Technology Center in São Jose dos Campos (*O Estado de São Paulo,* Dec. 9, 1983, p. 5). On the presumed nuclear weapons objectives of the "parallel program," see *Folha de São Paulo* (May 1, 1984, p. 9, and July 8, 1984, p. 23); *El Comercio* (Lima) (July 11, 1984, p. A2); and *O Estado de São Paulo* (Oct. 4, 1983, p. 10).

31. Such developments included material for control rods and high-purity zirconium for tubes (*O Globo,* Oct. 13, 1984, p. 16).

32. Components for the first research reactor were manufactured in Argentina (1958), except for electronic and control equipment (see Appendix C).

33. *El Desarrollo,* 1985, p. 178, and *Energeia* 29 (1985).

34. *El Desarrollo,* 1985, p. 170.

35. *Economic Information on Argentina,* Buenos Aires, 119, (1981): 13.

36. Cosentino (1984: Appendix 2). For instance, foreign personnel would perform managerial functions only with Argentine approval, thus ensuring local control of vital departments.

37. *Argentina Nuclear*, Apr.–May 1991, pp. 22–26.

38. Rodrigo et al. (1982).

39. *Nuclear Engineering International*, Feb. 1983; *Realidad Energética*, Apr. 8, 1983, p. 11.

40. Know-how in this area was even sold to an important German firm (Sábato 1973).

41. Sábato (1973:28); *Economic Information on Argentina* (Buenos Aires, 1982).

42. Kittl and Almagro (1982).

43. Reportedly this $62 million plant was to produce 500 kilograms per year of 20 percent enriched uranium by 1985 (*Energeia* 43, 1984, pp. 1203–4).

44. Redick (1978); *Nucleonics Week*, Nov. 7, 1985.

45. *Correio Brasiliense*, Mar. 24, 1975, pp. 4–5.

46. *Clarín*, Sept. 30, 1984.

47. Thome Filho (1983).

48. Biasi (1979).

49. CNEN, *Relatório Anual*, 1980, p. 142.

50. Tanis (1977); CNEA, "La dirección," n.d. Approximately 30 percent in the first category were high-level "independent researchers," 17 percent associated researchers, 34 percent junior researchers, and 18 percent research assistants (Tanis and Kittl 1976).

51. *Energeia* 47/8, 1984, pp. 17–19; CNEA, *Memoria Anual* (1981, 1982, 1983). CNEA's annual budget was estimated at an average of $1.5 billion to $1.8 billion between 1975 and 1983.

52. *Folha de São Paulo*, Apr. 28, 1978, p. 32. Silva (1984: Appendix 3) reports that Nuclebrás invested $23 million in "science and technology" by 1982. CDTN's budget also included allocations from FINEP and Banco do Brasil.

53. CNEN, *Relatório Anual* (1980). Its total budget for that year was about $33 million. The $3 million in R&D does not include an additional $7 million invested in training for the nuclear sector (almost one-fourth of its budget!).

54. Nuclebrás's total annual investments were estimated at about $800 million. These figures are a rough approximation, in light of the absence of data over all years. It is quite clear, however—from the National Plans for Scientific and Technological Development I and II—that R&D allocations within the nuclear sector decreased from 9.4 percent (1968) to 6.3 percent (1975–77), although not in real terms.

55. Fúnes (1984).

56. Nuclebrás, *O Brasil desenvolve*, n.d.

57. Nuclebrás, *Relatório Anual*, 1984, p. 10.

58. CDTN also established a testing laboratory for reactor components (Nuclebrás, *Pesquisa Nuclear no CDTN*, n.d.).

59. Tanis and Marrapodi (1985). The national coefficient for patenting across sectors was higher, about 31 percent. However, patenting in the nuclear sector requires higher levels of technological sophistication than other sectors, and in that sense, the ratio of 23 percent suggests significant achievements.

60. For a general overview of this literature, see Marton (1986).

61. Erber (1980).

62. INPI analyzed technical aspects, such as: Is the technology adequate? Can the buyer absorb it? Does the buyer have the appropriate human resources to do so? Upon its approval of a certificate (but no timetables), the firm registered the authorized ceiling for technology payments with the Central Bank. The Central Bank publishes all payments for technology, by sector, in its annual report, under Technical Assistance.

63. By 1978 INPI had approved a total of $468 million to be paid over ten years by Nuclebrás ($160 million), the electrical utility FURNAS ($284 million), and engineering and capital goods firms ($24 million) (*Relatório* 3, p. 156). INPI stood at the center of the highly charged political debate regarding the nuclear program (there were reports of formal investigations on INPI's deficient role), which may explain the reluctance of some INPI officials to be interviewed (even in the mid-1980s) in relation to the 1975 agreements.

64. Rosa et al. (1984).

65. This seems to have happened, for instance, in the case of an experienced construction firm, regarding an agreement with a German supplier for $25 million in technical assistance, or about one-fourth of the estimated value of the project (*Relatório* 6, no. 1, p. 462).

66. *O Estado de São Paulo*, Aug. 24, 1978.

67. In 1980 and 1981 there were a total (across sectors) of 499 technology transfer (licensing) agreements approved in the mechanical sector and 116 in the electrical and communications sector, with INPI controlling the terms, reducing restrictive practices, and avoiding excessive payments (Chudnovsky and Nagao 1983).

68. *Relatório* 3, p. 156. Contracts with 43 German suppliers of services and equipment were terminated in 1984 (*O Estado de São Paulo*, Nov. 18, 1984).

69. National Registry of Licensing and Technology Transfer Contracts (INTI), created by Decree No. 6.187/71 in 1971.

70. Jorge Sábato, an Argentine physicist, was a prolific writer on Latin American scientific and technological development who contributed to the fundamental conceptualization of the Argentine nuclear program. His approach is evident in the work of Sagasti (1978, 1979, 1980), Herrera (1972), Leite (1978), and others.

71. *Relatório* 3, p. 88.

72. By 1979 three university centers involved in state-supported nuclear

research were divorced from the program (*Relatório* 6, no. 3, p. 428). Neither were the relevant laboratories—such as the prestigious Institute of Technological Research and the Institute for Energy and Nuclear Research (IPEN) near São Paulo University (the former Institute of Atomic Engineering)—incorporated into the effort.

73. Batista opposed a 1974 proposal by the Ministry of the Army (approved by BNDES) for the construction of a heavy water facility at the Military Institute of Engineering (IME) in Rio (*Relatório* 6, no. 3, p. 466).

74. The institute had developed its own technology for production of uranium concentrate (yellowcake), pure uranium, and fuel elements (*Folha de São Paulo*, Jan. 5, 1978). By 1975 IPEN, Campinas University, and the Aerospace Technology Center were cooperating in the development of a laser enrichment technology (*Jornal do Brasil*, July 7, 1984). The 1975 agreements, instead, opted for an unproven technology, known as the jet-nozzle process. IPEN tested laser enrichment under laboratory conditions in 1979 (*O Estado de São Paulo*, Dec. 9, 1983).

75. *Jornal do Brasil*, Mar. 12, 1978.

76. Between 1970 and 1983 these institutions graduated 630 M.Sc.'s in nuclear-related disciplines (Thome Filho 1983).

77. Spitalnik et al. (1984).

78. *Relatório* 6, no. 5, pp. 174–77. According to some estimates, integration between academic research and the industrial program required between 50 and 60 new Ph.D.'s in physics per year (*Relatório* 3, p. 102).

79. The state's electrical utility Furnas and Coppe (a program at the University of Rio de Janeiro) began cooperating by gearing students' graduate theses to consequent employment and by procuring a model for the supervision of safety parameters and critical functions for Angra 1 (Thome Filho 1983).

80. CNEN offered fellowships to graduates of academic centers who were not being absorbed by the nuclear program and promoted agreements for basic and applied nuclear research with the federal universities of Pernambuco, Rio de Janeiro, Minas Gerais, Rio Grande do Sul, and Santa Catarina, and with other state agencies, such as the Aerospace Technological Center (at São José dos Campos), São Paulo's IPEN, the Institute of Nuclear Engineering (IEN), and Nuclebrás's CDTN in Belo Horizonte (*O Estado de São Paulo*, Jan. 12, 1986, p. 7, and Dec. 18, 1986, p. 29).

81. Tanis (1977).

82. Dahlman et al. (1985). Sagasti (1978) concluded that policies affecting industrial development more generally (i.e., trade, credit allocations)—or implicit instruments—seem to be more effective in strengthening technological development than explicit instruments such as science and technology plans, promotion of use of local technologies, fiscal incentives, and the like.

83. *World Bank Country Study* (1980:101). On FUNTEC's assistance to the capital goods industry, such as aircraft and computers, see Erber (1977).

84. It had supported about 50 percent of the nuclear engineers (M.Sc.'s and

Ph.D.'s) trained until 1973, prior to the nuclear agreements (Villela and Baer 1980:73).

85. These included the creation of EMBRAMEC at BNDES, nuclei of articulation between state firms and private industry, and FINAME's registration of local firms for the purposes of financing locally produced goods. EMBRAMEC (Mecânica Brasileira S.A.) was a new system of support for investments by the mechanical sector. BNDES strengthened its support for effective technology transfer by foreign partners, increased local execution of basic engineering projects as well as all detailed engineering, and emphasized its preference for national firms as main contractors, greater risk capital participation by foreign partners, and normalization of machinery and equipment specifications (*Relatório* 6, no. 5, p. 259).

86. BNDES, *Relatório de Atividades* (1981). Critics of the program suggested that BNDES should have taken part in the negotiation of nuclear agreements (*Relatório* 3, p. 102).

87. *Jornal de Tecnologia e Ciencia* (1978–79).

88. FINAME (Special Agency for Industrial Finance) subsidized financing for local capital goods purchases in conjunction with Banco do Brazil and EMBRAMEC. FINAME's financial programs served small- and medium-sized enterprises, capital goods in general, and custom-made capital goods (*World Bank Country Study* 1980:105). Major loan recipients were producers of equipment for steel, petrochemicals, railroads, and power generation and transmission. FINAME may have financed more than 50 percent of the equipment for Angra 2 and 3, as was conveyed to me during an informal interview at NUCLEN (May 1985), but I was unable to obtain any official confirmation of this amount. FINAME approved over $100 million for Nuclebrás in 1982, of which about $10 million was provided in 1982–83 and another $10 million in 1983–84. About $7 million had been offered in 1980–81 (see BNDES, *Annual Reports*, 1982, 1983, and 1984, and BNDES 1982b).

89. *Folha de São Paulo*, June 30, 1978; Nuclebrás, *Relatório Anual* (1977).

90. The fund was the major financial instrument of the Plan for Science and Technology (PBDCT). It channeled up to 16 percent of its allocations to new technologies, 60 percent of which went to nuclear energy (on average). This ratio preceded the 1975 agreements (Bond and Rao 1976:IV. 7–19).

91. Although there is some overlap between BNDES/FUNTEC and FINEP's activities, a division of labor allocated larger firms to the former and medium-sized and smaller ones to the latter. The overall performance of FINEP appears to have been mixed. In a comprehensive study of the private sector, Villela and Baer (1980) found that not many firms used FINEP's services, and those that did complained about its overbureaucratization and ineffectiveness. The same criticism was raised against INPI, which was perceived as overly concerned with legal barriers to royalty payments rather than with technological development.

92. *Folha de São Paulo*, June 30, 1978.

93. Personal interview at ABDIB (São Paulo, May 1985). Such was the case of Bardella and the technology for rolling bridges. In some cases licensing agreements seem to have been imposed as a result of pressures from NUCLEN's German-controlled technical committee (*O Estado de São Paulo*, Aug. 23, 1979).

94. Carvalho favored, for instance, mediation by a Brazilian research institute—such as the Institute of Technological Research or the University of Santa Catarina's Institute of Mechanical Engineering—in the evaluation of licensing requirements. These institutes were to be involved in the design of the containment wall, aiding Brazilian firms Montreal and Confab (Rosa et al. 1984:197).

95. Sábato (1973); Tanis (1977). From 1961 to 1967 SATI performed a total of 576 undertakings, mostly (475) in technical assistance to private industry, particularly in quality control, and under contract (101), in product or process development (Aráoz and Martínez-Vidal 1974:73). SATI's clients comprised about 130 private firms, 48 state or mixed capital enterprises, and 59 foreign firms.

96. This section builds on interviews with representatives from most major suppliers of components, engineering, and construction services. Interviews were based on a questionnaire I designed, partially based on another questionnaire used by Carl Dahlman and his colleagues for a separate study, which Mr. Dahlman kindly shared with me. Most interviews were conducted with technical personnel (senior managers) who, for the most part, insisted on anonymity, and occasionally with owners.

97. I interviewed five capital goods firms, four engineering firms, and one construction firm in Brazil. The five capital goods producers accounted, together, for over 80 percent of total Brazilian supplies of electromechanical components. Mechanical and electromechanical supplies account for a high share of total inputs for nuclear power plants. Brazilian firms contributed about 20 percent in this category for Angra 2 and 3. The two major engineering firms accounted for the bulk of Brazil's private engineering inputs, and for about 5 percent of the total engineering costs for the two plants. See Backhaus and Wildgruber (1982) for a more complete listing of participating companies and licensing agreements.

98. Construction and assembly engineering—unlike electrical machinery—are predominantly national sectors as well. Of 434 medium- and large-sized firms identified in 1981, 410 were in the hands of private local capital (*Estudios* 1982:25).

99. Each firm included between 100 and 200 engineers, and from 200 to 500 skilled technicians, roughly between 9.5 and 11 percent of the firm's personnel. The one (family-owned) medium-sized firm (274 employees) interviewed had a slightly higher percentage of skilled technical personnel (13 percent). Four (of the five) major participating capital goods firms were between 42 and 72 years old and employed between 1,500 and 7,000 people. The consortium of COBRASMA, CONFAB, and BARDELLA accounted for about 70 percent of domestic private supplies for Angra 2 and 3; the smaller

companies for the remaining 30 percent. The three are family enterprises owned by the Vidigal concern (the first two) and Bardella.

100. Product-design capacity in capital goods can be disaggregated into: feasibility study (determining whether a firm is able to produce the product defined by the client's needs); basic design (definition of the structure and major components of equipment, cost and time estimation); and detailed design (complete design for each part of the equipment, to allow manufacture of parts) (Chudnovsky and Nagao 1983:11–14).

101. Souza et al. (1972). Yet the medium-sized firm interviewed relied on in-house designs for 80 percent of its normal production. For the nuclear program, however, 100 percent was mandated to be of foreign (German) design. I found the average R&D expenditures among nuclear suppliers to have been below 1 percent of total sales, with one firm reporting 1–1.5 percent. In the U.S. the average percentage of R&D expenditures over sales was 2 percent for the capital goods sector (Chudnovsky and Nagao 1983:13). Only one firm had a specialized department for R&D, but all had active engineering or "trouble shooting" divisions conducting special testing and other adaptive, improving, and even developing, activities.

102. COBRASMA and CONFAB invested in the expansion of heavy components production ($80 million and $30 million, in Sumare and Pindamonhangaba, respectively).

103. The agreements were for ten years, no longer than many licensing agreements approved by South Korean agencies (Marton 1986).

104. Three firms performed process or product improvements on German designs, and two introduced material and product adaptations. Despite a generally higher ratio of imported materials for nuclear components than was true for other products (20 percent instead of 10 percent), these firms procured 80 percent of materials from domestic markets, particularly following a materials nationalization program in nuclear components introduced in 1979.

105. There were cases of improvements performed by Brazilian producers that were later transferred to the supplier.

106. One of the firms had produced 90 percent of the equipment fabricated in Brazil for Angra 1. Gains from Angra 1 in piping design, structural design, static and dynamic analysis, and process and electrical detailed engineering and geotechnology improved Brazilian firms' international competitiveness.

107. This was so because the firm had vast experience (longer than that of the technology supplier) in the production of conventional heat exchangers and had only to assimilate nuclear specifications.

108. These improvements were reflected in production techniques, plant layout, quality of materials and parts, operating and maintenance procedures, tooling usage, testing of products and materials, and labor skills, and the assimilation of new types of stress analysis (for instance, in the production of the containment vessel). Metrology, calibration, reliability, and nondestructive testing were perhaps the greatest beneficiaries of the introduction of new quality assurance and quality control standards.

109. *Relatório* 6, no. 3, p. 121. The manufacturing of very large systems often requires different techniques of handling and machining as well as the use of new equipment. Size, therefore, may involve a higher degree of technological complexity (Teitel 1984).

110. Despite the general absence of formal agreements between private firms and local research institutions, some producers did turn to the Institute for Technological Research (IPT), among others, for consultations and testing. Such cooperation resulted in new developments by at least two firms (the expansion joints for heat exchangers and clad steel). A third firm had an agreement with the University of São Paulo for the development of computer programs for calculations regarding specific components. A fourth consulted with CEMPES and IPT for measurement of instrument reliability, and a fifth with IPT and CEPEL for mechanical tests.

111. My survey of Argentine firms included two major suppliers of components and two major engineering firms. One supplier controlled about 50 percent of the market for processing equipment and pressure vessels and 70 percent of some hydroelectrical components (turbines, generators).

112. Private firms signed a total of between 22 and 30 technology transfer contracts in the nuclear sector until 1984 (as opposed to 180 in metallurgy, 167 in oil, 138 in the chemical sector, and 125 in the hydroelectrical sector) (Tanis 1985:29; *El Desarrollo*, 1985, p. 186). Royalties for nuclear-related licenses reached, at times, 15 percent of sales.

113. For a list of the major industrial firms involved in the Argentine nuclear industry, see Tanis and Ramberg (1990:96) and *Eye on Supply* (Monterey Institute of International Studies), no. 1 (Summer 1990): 55–60.

114. *Economic Information on Argentina* 13, no. 128 (1983).

115. The Argentine firm had wide experience in production for nuclear power plants and experimental facilities, the gas and oil industry, steel mills, and other projects (Pescarmona 1982).

116. See *El Desarrollo*, 1985, pp. 146 and 186.

117. Sábato et al. (1978:94–100); *La Prensa*, Jan. 9, 1984. One firm developed a special welding procedure that would arguably allow domestic production of all heavy components for nuclear plants (*El Desarrollo*, 1985, p. 187). On gains in the areas of special concrete, special alloys, and quality control, see *Clarín*, June 1, 1984.

118. Tanis and Marrapodi (1985:11).

119. *El Desarrollo*, 1985, p. 58.

120. The four Brazilian engineering firms interviewed were all between 18 and 30 years old. Each had between two thousand to three thousand employees, 25 percent of whom had engineering degrees, and a large number of technicians with some degree of formal training. Two had full-fledged R&D departments, rendering outside services to private and public firms. One of the firms spent between 1 percent and 3 percent of its sales in R&D, up until 1980, when macroeconomic conditions forced a retrenchment. Another devoted 22 percent of its personnel costs to feasibility studies and technology

(basic and process design). At least two firms had developed new processes and products (particularly in other energy-related areas) and held patents.

121. Engineering firms in Brazil routinely perform detail engineering and purchase equipment and basic designs abroad (Ford et al. 1977).

122. Involvement with the nuclear program represented from 5 percent to 25 percent of total revenues for engineering firms. Eight of the most important firms in the country performed piping and ventilation engineering services (Rosa et al. 1984:268). Natron also supplied services for Nuclebrás's yellowcake plant, at the Poços de Caldas Industrial Complex. On Promon's technological investments for the nuclear program, see Muller et al. (1980). On its prior experience with Angra 1, see Venancio et al. (1980).

123. Private firms estimated, for instance, that they could have performed about 90 percent of all engineering services for a nuclear plant, particularly building on their prior involvement with Angra 1 (Muller et al. 1980). Former NUCLEN official Joaquim Carvalho claimed private Brazilian engineering firms could have performed most of NUCLEN's undertakings at a lower social cost (*O Estado de São Paulo*, June 7, 1985).

124. The most important gains included in improved capacity in dynamic analysis, stress analysis, modeling, and simulations, as well as gains in the selection of computer software for calculations, where in-house development was emphasized. There was at least one case of product development related to the nuclear program.

125. These firms cooperated routinely (in nonnuclear projects) with universities (São Paulo and Rio de Janeiro) and other local research agencies (such as Petrobrás's CENPES) in the development, building, and testing of equipment and processes.

126. A consortium of firms involved in the nuclear program provided detail engineering, construction supervision, and assembly services for a Peruvian reactor (*Clarín*, June 1, 1984).

127. *FBIS*, Latin America, July 16, 1984. Another firm, with no prior export experience, became an international supplier (of condensers piping), and yet another was involved in the sale of a radioisotope facility to Algeria (*Nucleonics Week*, Nov. 28, 1985).

128. Argentina's overall performance in project exports was not considered to be very impressive in comparison to other NICs (Dahlman and Sercovich 1984).

129. In the Peruvian case, for instance, there were considerable difficulties involving both Argentine financing and Peruvian budgetary constraints.

130. On these differences, see Katz (1985:133).

131. CNEA was able, for the most part, to shelter the program from the country's financial upheavals until the late 1980s. Only the emergence of a privatizing consensus in the early 1990s dealt it a deadly blow.

132. It is important to bear in mind that many of these Argentine firms had been dedicated to the nuclear sector (in fact, quite a few effectively took off in connection to the sector's growth), in contrast to their more diversified Brazilian counterparts.

133. Some of this expertise was transferred to a conscientious effort to develop national technologies and design capabilities, through the $1–3 billion parallel program (*EFE*, Apr. 22, 1987).

134. On technological development in Latin America as a by-product of manufacturing activities, rather than of dedicated R&D efforts, see Teitel (1981). On the importance of production experience in maintaining industrial competitiveness, see S. Cohen and Zysman (1987).

135. Engineering firms, for instance, had used sophisticated tools such as computer-aided design for conventional power plants.

136. Neither does the evidence support a traditional association of Latin American industrialists with lack of technical or organizational ability and interest (Lipset and Solari 1967). Given the program's strategic uncertainty in Brazil and the even greater macroeconomic uncertainty in Argentina, private entrepreneurial participation seemed more in line with generic patterns of firm behavior.

137. Mytelka (1979). On how domestic learning, often through state enterprises, lubricates the "obsolescing bargain" (or the decline in foreign investors' bargaining advantages), see Moran (1974). On how successful domestic (often family-owned) conglomerates in Brazil and South Korea (*grupos* and *chaebols* like Hyundai, Samsung, and Daewoo) relied on licensing, imported know-how, engineers trained abroad, and imported capital goods, see Encarnation (1989:17).

138. *El Desarrollo*, 1985, p. 99.

139. On advantages and disadvantages of licensing, see Dahlman et al. (1985) and Frame (1983).

140. Dahlman et al. (1985). Angra 1, supplied by Westinghouse to Brazil before the German agreement, supports that case as well. Although contractually defined as a black box, Angra 1 became an effective learning experience (Rosa et al. 1984:249, 289).

141. On India and South Korea, see Encarnation (1989:18).

142. Dahlman et al. (1985). I have not explored this particular angle here, but it is worthy to note that Brazilian and Argentine firms concurred in their claims that German suppliers were generally more reluctant than U.S. and other European firms to transfer procedures, standards, and conceptual designs and to subcontract locally.

Chapter Seven

1. Brazil's Minister of Mines and Energy Shigeaki Ueki (quoted in *Manchete*, Apr. 24, 1976, pp. 75–97).

2. Colonel Luiz Francisco Ferreira, former Advisor for Nuclear Affairs at the Ministry of Mines and Energy, in his deposition to the Parliamentary Investigating Committee on the Nuclear Program (*Relatório* 6, no. 3, p. 121).

3. Even leading critics of the characteristics of Brazil's program adopted in 1975 held such beliefs. See, for instance, José Goldemberg, *O Estado de São Paulo*, Apr. 22, 1979, p. 22 and *Opinião*, July 18, 1975, p. 8.

4. On these concepts, see Hirschman (1967).

5. Spin-offs and linkages may overlap to some extent. However, not every new skill or material, or improvements thereon, is accompanied by industrial-scale applications. A useful analytical distinction (that may be blurred in empirical studies) is to approach spin-offs as a technological concept and linkages as a commercial one.

6. On spillover benefits of high technology to the broader economy, see Tyson (1993:32–52), and Cohen and Zysman (1987).

7. *NASA Spin-off* (Introduction—all volumes). Spin-offs and spillover effects are used interchangeably throughout this chapter. In the U.S., NASA was responsible, by Congressional mandate, for promoting further use of the technology it developed. The instrument created was the Technology Utilization Program, which oversaw technology transfer from its own bank to potential users.

8. As Rosenberg (1982:71) argues, "Many of the benefits of increased productivity flowing from an innovation are captured in industries other than the one in which the innovation was made."

9. Parker (1978:121).

10. Trajtenberg (1983:14) studies *innovative* activity only, but his argument can apply to technological modernization more generally.

11. Erber (1981b).

12. Weston (1986).

13. Some of these categories were inspired by Weston (1986).

14. Rosenberg (1982).

15. Nuclear components such as heat exchangers, pressurizers, and pressure vessels require the use of specialty steels (austenitic, high-nickel content, and others) with large cross-sectional dimensions and high strength (Sábato et al. 1978). Some components in the auxiliary systems also require special materials and processes (welding, thermal treatment, founding). Metallurgy problems range from the purity of the uranium dioxide used as a fuel, to the mechanical properties of the pressure vessel, fuel element manufacture, corrosion-resistance of heat-exchange tubes, basic physical properties of different metals, alloys, and oxides; and techniques in foundry, forging, rolling, extrusion, and welding (Aráoz and Martínez-Vidal 1974).

16. Rosenberg (1982).

17. Joaquim Carvalho (personal interview, May 1985).

18. A type WStE-51 steel, imported for the first plant (Angra 2), was to be substituted for the WStE-370 type (Nuclebrás, *Relatório 1984*, p. 10). Yet all stainless steel for both plants had to be imported (from German sources) despite the local availability of an almost identical type with similar specifications (private interview, Rio de Janeiro, May 1985).

19. NUCLEP procured ten thousand tons of metal structures from the National Steel Company (CSN) in 1977 (*Jornal do Brasil*, Jan. 25, 1977, p. 4). National steel companies learned to produce new kinds of special steels, despite KWU's reluctance to nationalize steel inputs.

20. This included geological prospecting techniques, cartography, magnetic measurements, and aerographic surveys (Carvalho 1981).

21. One research institute developed a new technique to exploit uranium (by means of a local bacteria capable of oxidizing iron) and transferred it to Nuclebrás (*O Globo*, Nov. 12, 1978, p. 42).

22. *O Estado de São Paulo*, Oct. 22, 1983.

23. These products included cerium oxide and fluoride, lithium hydroxide and beryllium oxide, ilmenite (used in abrasives, electrodes, and iron alloys) and rutile (used in electrodes for electric welding, pigments, and in ceramics) (*Brazil Energy*, June 8, 1983; *O Globo*, Oct. 13, 1984, p. 16). NUCLEMON, created independently from the 1975 agreements, also developed a technique to obtain high-grade neodymium oxide (99.9 percent), while the University of Campinas fabricated the corresponding crystal and used it successfully in lasers (Nuclebrás, *Relatório 1984*, p. 7).

24. The production of calcium molybdate was expected to substitute for the importation of molybdenum trioxide for the steel industry (*O Estado de São Paulo*, June 25, 1983, p. 25).

25. The materials and products resulting from these activities, as well as the skills and processes themselves, can be applied in other industrial areas, including energy, chemical and petrochemical, aerospace, metal-machining, and transportation.

26. Nuclebrás, *O avanço*, n.d.

27. A Nuclebrás prospectus reports service contracts with 150 firms and institutions, private and public, such as ALUMAR, Salgema, CEMIG, Portobrás, Petrobrás, Vale do Rio Doce, and others, in chemical and pressure analysis, quality control, destructive and nondestructive testing, welding, and testing of mechanical, electrical, and electronic components (Nuclebrás, *O Brasil desenvolve*, n.d.).

28. These included a new type of refractory material to be used in aluminum melting and a new process for the manufacture of aluminum evaporators for refrigerators, among others (Sábato 1973).

29. The electromechanical metallurgical industry already represented 25 percent of all Argentine industry at the time and in GNP terms was more important than the agricultural sector (30 percent to 25 percent) (Sábato 1973).

30. CNEA aided industrial firms in adopting processes in the area of high-strength weldable steels (Tanis 1977:76–80). In 1968 an industrial firm developed the first niobium steel alloy. These techniques were later incorporated in the production of large-diameter bars for concrete. CNEA's steel and rolling experts provided assistance in streamlining production for transmission lines in the hydroelectric projects of El Chocón and Salto Grande. Finally, it developed a procedure to harden soft woods (through irradiation), applicable in the construction industry.

31. Aráoz and Martínez-Vidal (1974).

32. At the request of an industrial firm producing nonferrous material,

CNEA developed a material purification technique that was later patented (Tanis 1977). Research applied to processes of chlorination led to several patents as well. Other new materials and techniques included a new metallographic technique for the nondestructive testing of boiler tubes; the application of radioisotope techniques to study the weight of steel balls used in mineral grinding; the measurement of liquid levels in industrial tanks; and the use of fracture mechanic techniques to evaluate the probable life of pressure vessels (Sábato 1973; Tanis et al. 1980).

33. The construction of nuclear power plants requires design and construction of steam generators, recuperative feedwater heating systems, pumps, closed cooling, air conditioning, water treatment systems, heat exchangers, pressure vessels, tanks, and other equipment. Mechanical parts account for about 40 percent of supplies for a nuclear plant.

34. On technological convergence in the capital goods sector, see Rosenberg (1976 and 1982). The specialized structure of production in capital goods, particularly custom-made capital goods, requires suppliers of subcomponents, parts, tools, and raw materials to adjust to new specifications, thus expanding technical knowledge and experience (Chudnovsky and Nagao 1983).

35. *Relatório* 5, p. 40. NUCLEP built metal structures for submerged platforms in the Cherne and Namorado oil field for two private firms (*O Estado de São Paulo*, Oct. 23, 1983, p. 7). The Navy contracted for NUCLEP's services in the construction of HVW submarines (*Folha de São Paulo*, Aug. 24, 1986). The attempt to use NUCLEP's idle capacity in the fabrication of Leopard war tanks, used by NATO, failed following military opposition apprehensive that NUCLEP might compete with the Brazilian firm Engesa, a successful producer of military vehicles.

36. Joaquim Carvalho (personal interview, May 12, 1985). The design of the reactor core involves the use of computer design codes of high accuracy.

37. Castro (1974).

38. On standardization in the nuclear field and technological modernization, see Tychojkij (1982).

39. In one instance, CNEN rejected a KWU design of a portion of the structure for a power plant, and KWU was forced to revise the depth and number of pilings supporting the reactor building (*O Estado de São Paulo*, Oct. 20, 1983, p. 9).

40. Professor José Ribeiro da Costa, vice-president of Brazil's Committee on Nuclear Energy (personal interview, Rio de Janeiro, May 1985). Professor da Costa taught Nuclebrás's first course on principles of quality assurance in 1974 to 126 engineers, physicists, and professors from Petrobrás, Westinghouse, Aerospace Technology Center (EMBRAER), private industry, and laboratories. Quality assurance encompasses formalization of programs, methods, training, design and engineering, manufacture, assembly, and operation activities. Quality control refers to the testing of the final product for potential deficiencies.

41. The German firm Rheinischer-Westfalischer Technischer-Verein

(RW-TÜV) provided technical supervision, training, and technical assistance (*Relatório* 3, p. 156).

42. Personal interview (IBQN, May 1985). IBQN provided services for over 60 firms, in nuclear and nonnuclear areas. It expanded the use of techniques such as nondestructive testing, involving inspection of materials, products, and equipment, without altering their capacity to be used.

43. A single major component producer required about ten subsuppliers of materials and fifteen subsuppliers of components. Certification was required in areas such as forging and casting (heat processes), bars, tubes, screws and bolts, electrodes, and copper tubes.

44. Knodler (1982).

45. *Relatório* 6, no. 3, p. 115.

46. On product improvements and cost reductions among industrial purchasers of capital goods as a result of upgraded production of capital goods, see Rosenberg (1982:76).

47. Experience gained with Angra 1 in quality assurance was applied to high-voltage (direct current) transmission lines, hydroelectrical, oil, and subway systems.

48. Other firms applied norms and standards used in the production of special filters for water treatment in the nuclear circuit to other industrial areas. New ultrasonic and X-ray techniques (for nondestructive testing) were applied to the production of offshore platforms. Engineering firms used the experience gained in seismic design for offshore platforms and hydroelectrical plants. In general, however, few areas require the same level of sophistication in seismic analysis that a nuclear program does.

49. Petrobrás started its own quality assurance program in 1981, about six years after the nuclear program.

50. Rosa et al. (1984:199). The strong capability among suppliers of the aerospace industry (for instance, in casting processes), preceded production for the nuclear sector.

51. About 71 firms (in metallurgical components, boilers, engineering and assembly, services, and shipyards) employ personnel certified in industrial radiography, penetrating fluids, magnetic particles and ultrasonic techniques (Baez et al. n.d.).

52. Tanis et al. (1980). On CNEA's activities in standardization and quality assurance, see Tanis (1977:77–80).

53. *FBIS*, June 7, 1984, p. 16. Another group of CNEA scientists was transferred to a state firm, High Technology (Altec), established in the microelectronics sector (*FBIS*, June 13, 1985, pp. 5–6).

54. Martínez Nogueira (1978). On the impact of experience gathered in the nuclear program for the disaggregation of SEGBA's Machine IV and the hydroelectric project Yaciretá, see *El Desarrollo*, 1985, p. 168.

55. Sábato et al. (1978). The standards of the nuclear industry had some, albeit limited, effects on the chemical, naval, and electrical industries (Tanis 1985).

56. Hirschman (1967). Trained personnel played a critical role in the de-

velopment of nuclear industry in advanced countries, where the state assumed that responsibility initially (Kuhn 1966).

57. Vargas (1976). According to a 1975 estimate, the nuclear program (including Nuclebrás and subsidiaries, the Atomic Energy Commission, and private firms) would require 4,500 new university graduates and 5,500 new technicians (a total of 10,000) until 1985 (Alves et al. 1977:11). In 1982 the estimate decreased to 4,250 (until 1985) (*Relatório* 5, p. 34).

58. Spitalnik et al. (1984:24); Spitalnik and Lerner (1984:7).

59. This projection is from Alves et al. (1977) and Nuclebrás's *Relatório 1977*. Ph.D.'s were to constitute about 20 percent of trainees across most sectors (Spitalnik 1982:19). While the program forecast the training of 258 M.Sc.'s and 154 Ph.D.'s by 1978, only 84 percent and 3 percent, respectively, graduated up to then (*Relatório* 6, no. 5, p. 174).

60. Lichtenberg (1985).

61. *Boletím S.B.F.* 5, (Year 6) (1975). Up to 1980, 678 fellowships were granted in nuclear engineering, 9 in nuclear physics, 19 in geology and mineralogy, and 26 in nuclear chemistry. Brazil's Physics Society suggested the need for 2,500 high-level scientists (including 620 physicists, two-thirds M.Sc.'s and one-third Ph.D.'s) for a more self-reliant program, and 1,400 high-level scientists (including 350 physicists) for one aimed at absorbing foreign technology (*Boletím S.B.F.* 6, no. p. 9). A study of staffing in the nuclear sector in the United States suggests that the proportion of mathematicians and physicists in a nuclear program was as important at later stages as during R&D (Kuhn 1966:72).

62. *Relatório* 4, p. 202; Fabricio (1982:11).

63. It also ran a center for youngsters recruited in the area (Itaguai); these youngsters spent about three years training in welding, tools, assembly, and other tasks. (On training figures at NUCLEP, see *Relatório* 4, p. 209.)

64. Fabricio (1982); Spitalnik et al. (1984). Some regarded this number as insufficient, including the German Minister of Energy (*Relatório* 6, no. 1, p. 327). For the sake of comparison, India had sent close to twenty thousand technicians to train in Canada on transfer of CANDU reactor technology.

65. Nuclebrás, *Relatório 1984*, p. 12. KWU required between $5,000 and $30,000 per trainee as a "matriculation fee" (*Relatório* 6, no. 3, p. 447; Rosa et al. 1984:214). The capital cost per trainee per month was $3,700 (personal interview, NUCLEN, May 1985; *Relatório* 6, no. 5, pp. 496, 503). Public reports of total costs per trainee—over the length of training—were as high as $400,000 (*O Estado de São Paulo*, Nov. 11, 1984).

66. *Relatório* 6, no. 5, p. 496; Leite (1978); *O Estado de São Paulo*, Oct. 20, 1983, p. 9. On the limited transfer of technical knowledge, see the deposition by Professor Milton Campos, an expert in training in the nuclear field (*Relatório* 6, no. 4, pp. 380–81). The figure of 58 people trained in R&D amounts to between 10 percent and 15 percent of those trained abroad (Fabricio 1982:22).

67. Spitalnik et al. (1984:25). Both the transfer of contractual scope respon-

sibility for the design of Angra 2 and 3, and the reduction in managerial positions held by Germans at NUCLEN (from 93 percent in 1977 to 28 percent in 1982), were also considered an outcome of successful training (Lepecki 1985; Fabricio 1982).

68. Erber (1981a); Landes (1992).

69. Among the academics, 60 percent were engineers and 15 percent geologists. Among the medium-level professionals, 62 percent were technicians, 10 percent projects designers, and 9 percent welders. About 20 percent of a group of 158 engineers trained in Germany (in heat exchangers, steam generators, quality assurance and control, mechanical and electrical equipment, materials technology, automation, and instrumentation and control) left NUCLEN at that time (*O Estado de São Paulo*, Nov. 18, 1984, p. 43).

70. Spitalnik and Fonseca (1985). On the potential contribution of trainees to other energy programs, see Joaquim Carvalho (*O Estado de São Paulo*, June 7, 1986).

71. There were private costs as well. A single producer lost 30 percent of its trainees for the nuclear program to other capital goods firms not involved in it. About 350 engineers were reportedly laid off by three major engineering firms involved in the program, once the program was scaled down in the early 1980s (*O Estado de São Paulo*, Oct. 19, 1983).

72. By 1978 there was a total of 223 graduates from quality assurance courses, over 50 percent from the private sector, and the rest from state-owned enterprises. Many had prior experience with quality *control* techniques, but only about 30 percent had experience with applied quality *assurance* techniques up to 1978. That year marked the beginning of a rapid diffusion of these concepts. Within Nuclebrás alone, about 200 were trained specifically in quality assurance and quality control courses between 1974 and 1981, and a total of 1,300 participated in short courses on the subject (Spitalnik 1982:22).

73. Sábato (1973:25).

74. Tanis (1977:54) and Tanis et al. (1980). CNEA created the first academic course in Physics of Metals at its own Institute Balseiro. Among the courses taught within CNEA were Methodology and Radioisotope Applications (about one thousand graduates), Radiotherapy and Dosimetry, and Metallurgy. CNEA's acronym was jokingly interpreted as standing for Comité Nacional de Educación Atómica (National Atomic Education, rather than Energy, Commission), thus singling out its commitment to training and education.

75. Tanis (1977). By 1972, CNEA's scientific and technical personnel of 920 amounted to 31 percent of 2,970 employees (500 with Ph.D.'s or equivalents). These proportions were equivalent to those of the U.S. during the development of its nuclear industry (Kuhn 1966:73).

76. *Relatório* 6, no. 6, p. 107. Some of the depositions to the parliamentary investigating committee blamed Nuclebrás's own efforts to discourage pub-

lications, as well as the agency's excessive secrecy, for this poor record. The relaxation of political and security concerns and a relative opening of the nuclear program to public debate in the early 1980s brought about considerable changes in this area (Rosa et al. 1984).

77. Sábato (1973:37). It had trained about 1,000 people abroad (and hosted 350 foreign experts) by 1978, four years after completion of the first plant (Atucha 1) (Sábato et al. 1978:4; *Energeia*, Aug. 1984, p. 40). Among those were lathe operators, milling machine operators, die makers, glass blowers, carpenters, microscopists, surveyors, and cartographers. Overall, about 130 metallurgists were trained abroad by 1973 (Sábato 1973).

78. *Realidad Energética*, Buenos Aires, Mar. 1984, p. 28.

79. Personal interviews, Buenos Aires, May 1984.

80. Tanis (n.d.:374 and 1977:55). Over two hundred (37 percent of them foreigners) graduated from courses in metallurgy.

81. Fishlow (1971). Ranis (1984:100) defines "compulsive sequences" as "new ideas at one processing stage forcing appropriate changes in techniques and attributes up or down the line."

82. Some of the categories in Figure 4 were inspired by Fishlow's (1971) study of the introduction of American railroads.

83. O'Donnell (1978) refers to backward linkages as "deepening." See also Kaufman (1979) and Serra (1979).

84. It would be inappropriate, for instance, to infer the impact of the nuclear program on backward-linking industries through the rate of growth in demand for their products, since an array of other factors operated concomitantly, which might have influenced that rate. To the best of my knowledge, there is also no study of linkages of nuclear programs for developed countries. It is possible that nuclear industries in the industrialized world benefited more from the development of other sectors (such as computers), rather than the other way around. On how studies of foreign investment have not yet determined definitive ways to assess backward-linkage effects, see Grieco (1986).

85. On Siderbrás's plans to use NUCLEP's port, see *O Estado de São Paulo*, July 30, 1983.

86. *Gazeta Mercantil*, Oct. 21–22, 1984, p. 11. Some capital goods firms in the area of radioactive minerals prospecting and research benefited from the substitution of imports in well-logging and other equipment.

87. The careful selection of components to be produced locally was based on an assessment of their potential for affecting converging or input-supplying industries. On this aspect of CNEA's strategy, see *El Desarrollo*, 1985, pp. 45–132.

88. See article by former CNEA scientist Amílcar J. Funes in *Somos*, Apr. 27, 1984.

89. On Brazil's energy balance, see Erickson (1981a and 1981b) and Solingen (1991). The availability of alternative energy sources was a critical component of the debate between advocates and critics of the nuclear pro-

gram. The effects of Brazil's nuclear-generated electricity on cost reduction in energy rates and reliability of supply are hard to assess; none of the plants planned by the 1975 agreement produced electricity as of 1995.

90. In defining the three major phases of Brazil's industrial development, the president of the capital goods producers' association (ABDIB), Silvio Puppo, pointed to the implantation of a national steel industry (after World War II), the automobile industry, and the introduction of a nuclear industry. The last one, he argued, would enable producers to supply any other sector in light of its stringent quality assurance and safety requirements (Rosa et al. 1984). Although exports might not be conceived of as forward linkages in a strict sense (they do not necessarily imply the establishment of new industries but an *expansion* of an existing one), I include the foreign demand for nuclear equipment and services as forward linkages here.

91. See, for instance, President Sarney's speech at the Planalto Palace (Brasília, Sept. 4, 1987). CNEN, the national Nuclear Energy Commission, had been interested in exporting technology for licensing nuclear installations since 1969.

92. On technology exports as an indicator of changes in comparative advantage resulting from industrialization, see Dahlman and Sercovich (1984). There can be a "learning by teaching" effect to technological exports as well. The term is quoted in Aráoz and Sercovich (1977) and is attributed to Mario Kamenetzky.

93. *World Bank Country Study* (1980:109). On the emphasis on export promotion after 1964, see Hirschman (1981).

94. Nuclebrás, *Relatório 1984*, p. 8; BID/INTAL, 1983, p. 73; *FBIS*, Feb. 9, 1984, pp. 36−37. Bids for Peru and Mexico were canceled when these countries slowed down their nuclear programs (*O Estado de São Paulo*, Oct. 23, 1983). For an overview of Brazil's nuclear exports, see Solingen (1990a).

95. KWU arranged NUCLEP's export to Argentina (a client of Siemens and KWU). Nuclebrás had to agree to KWU's requirement that NUCLEP approve exports of nuclear material or services in Latin America (*Brasil Energy*, Oct. 12, 1981, p. 10). KWU and its associates granted export licenses for nuclear reactors and fuel elements, and for uranium enrichment facilities, hardware and blueprints of reprocessing installations (*O Estado de São Paulo*, Oct. 24, 1978).

96. According to Artur Gasparian, from a major engineering firm subconstracted by NUCLEN, by the early 1980s Brazil was capable of exporting engineering services for nuclear plants in areas such as site selection, structural design (including special seismic analysis), special piping, complete design of the nonnuclear portion of nuclear plants, and field engineering (Rosa et al. 1984:221).

97. *Brazilian News Agency*, Feb. 15, 1985. One source estimated Brazil's uranium export capacity to be sufficient to supply 48 reactors (1,300 MW) during 30 years (*O Globo*, Oct. 29, 1983, p. 16). Financial difficulties precluded Nuclebrás from selling 370 tons ($22.2 million) per year (1986−89) to

a U.S. firm and 150 tons per year (1988–2003) to Turkey (*O Estado de São Paulo*, Nov. 6, 1984, p. 34; *Gazeta Mercantil*, Nov. 6, 1984, p. 13.

98. *O Estado de São Paulo*, Jan. 14, 1984, p. 22. It also exported beryllium oxide to the United States in 1985.

99. Nuclebrás liaised between the engineering and construction firms, offering its own laboratories to define technical parameters, equipment, and the like (*Folha de São Paulo*, Sept. 27, 1984, p. 10). The bid was in the amount of $60 million, and civil works were budgeted at about $300 million.

100. Redick (1981); *Jornal do Brasil*, Sept. 27, 1979. On low external demand for enriched uranium, see *Relatório* 6, no. 3, p. 415. On the price disadvantage of Brazil's enriched uranium, see *Relatório* 6, no. 6, p. 66.

101. *Gazeta Mercantil*, Nov. 6, 1984, p. 14.

102. *Estudios*, 1982, p. 33.

103. The firm was responsible for the supply of 74 heat exchangers and two boilers for Angra 2 and 3.

104. Dahlman and Sercovich (1984).

105. On Promon's experience with Angra 1 (supplied 49 percent of conventional engineering performed in Brazil) and its exports record, see Gasparian and Calvet (1979).

106. From 1966 to 1979 the share of capital goods exports increased from 1.8 percent to 14.7 percent of total exports, including transportation equipment and nonelectrical machinery. Capital goods exports (together with construction) already accounted for most of Brazil's technology exports by the 1980s (Dahlman and Sercovich 1984). Custom-built capital goods exports amounted to $800 million in 1983 (*ABDIB Informa* 194:3). On Brazil's comparative advantage in mechanical machinery, see Fajnsylber (1971) and Mitra (1979). Exports of engineering services took off in the mid-1970s, prior to the nuclear program.

107. Redick (1981). For further details on the Argentine-Algerian connection, see *Bulletin*, Emerging Nuclear Suppliers and Nonproliferation Project (Monterey Institute of International Studies), Apr. 16, 1991.

108. *Nucleonics Week*, Nov. 28, 1985, p. 7.

109. *Clarín* Suplemento Especial, June 1, 1984, p. 4.

110. Mexico, Bolivia, and Chile, all with extensive experience in extractive metallurgy, sought training in the metallurgy of transformation (Tanis et al. 1980:18).

111. *Nucleonics Week*, Nov. 28, 1985, p. 7. Similar services were reportedly provided to Algeria. CNEA sent experts—under IAEA and OAS sponsorship—in general prospecting, airborne prospecting techniques, geochemical and geophysical techniques, and analytical and mineralogical lab techniques, as well as in recognition and evaluation of uranium ore-bearing areas, and uranium mine and mill projects and operation (Rodrigo et al. 1982). Other agreements were signed with Rumania and Turkey (Tanis 1985: 9). A Czechoslovakian minister visiting Argentina in late 1985 suggested that the two nations explore joint marketing of Argentine nuclear parts and

technology in South America and Europe, including COMECON countries. The Czechoslovak's main interest, when offering joint development of Argentine uranium reserves and fuel-cycle technology, was in selling heavy mining equipment and nuclear station components to Argentina.

112. *FBIS*, May 16, 1994, p. 10.

113. These developments renewed nonproliferation concerns at the time, in light of the effective absence of both countries from the Nonproliferation (NPT) and Tlateloloco Treaties until the early 1990s (Redick 1981). On Brazilian-Argentine positions regarding the NPT, see Redick (1988).

114. On the international strategic aspects of nuclear exports, see Solingen (1990a).

115. On the advantages of this strategy, see Hirschman (1967).

116. On the limited extant research on the political economy of skill formation, see King (1984).

117. Casting, forging, machining, heat treatment, assembly, and testing skills may be acquired on-the-job or through vocational training, and their diffusion in the economy is of considerable importance (Chudnovsky and Nagao 1983: 10). On the overskilling of labor, see Felix (1974).

118. Edmar Bacha is credited with coining the term *Belindia*.

119. *Clarín*, Apr. 13, 1984, p. 14. Alfonsín's Secretary of Foreign Affairs, Jorge F. Sábato, argued that, despite some limited contributions, the nuclear program had no recognizable "dynamizing" influence on the country's economy (*Clarín*, Sept. 22, 1984, p. 10). See also *Clarín*, May 19, 1985. On technological diffusion in Argentine industry, see Caputo (1978: 11).

120. On hybrid agencies, see Hirschman (1967). On the negligible participation of Córdoba's strong mechanical sector in the nuclear industry, see *Energeia*, 1985, p. 25.

121. A study by Villela and Baer (1980) shows that 90 percent of private industrial firms in Brazil are small- (under one hundred employees) or medium-sized. In addition, the firms interviewed were producing mainly custom-made items (such as complex boilers and heat exchangers) rather than products resulting from continuous-flow processes.

122. Over 980 Argentine firms provided automobile parts for the domestic market and exports, employing 130,000 people by 1974 in manufacturing, 146,000 in distribution and commercialization, and 187,000 in related services, for a total of 463,000 (Sourrouille 1980:151–63). In Brazil, about 140,000 people worked in production of cars and parts as early as 1960, and a total of 700,000 were employed in related sectors, including 1,200 producers of parts (Gattas 1981:326).

123. Many capital goods and engineering firms interviewed expressed the view that the oil and petrochemical sector in Brazil had much broader multiplier effects, technologically, than the nuclear industry.

124. In training, for instance, many of the engineers trained by the nuclear program had prior academic degrees in mechanical, electronics, metallurgical, or chemical engineering. Their conversion to the nuclear field prevented

their contribution to other fields. Private firms hiring very specialized engineers (up to 50 in one firm alone) misused their resources.

125. Carvalho and Goldemberg (1980).

126. Reppy (1985).

Chapter Eight

1. Technological decisions are often the outcome of the political nature of mutual adjustments among state and private actors. On the lack of systematic *comparative* research on scientific and technological efforts in different socioeconomic contexts as one of the most obvious deficiencies in science policy studies, see Spiegel-Rösing and Solla Price (1977).

2. On high macropolitical consensus and on the hegemony of the Economic Planning Board in South Korea, see Gereffi (1990:98) and Chang (1994:125–26).

3. The Tucuruí hydroelectrical plant, for instance, was designed to provide subsidized electricity, among other targets, to the aluminum plant established through a joint venture between Cia. Vale do Rio Doce and a Japanese concern (Mirow 1979:203).

4. On the similar impact of high segmentation (that is, low sectoral autonomy) in Mexico's nuclear program, see Stevis and Mumme (1991).

5. Grieco (1984); Encarnation (1989).

6. Brazil's relatively higher macropolitical consensus stemmed from the fact that all critical partners in the ruling coalition perceived the model implanted in the late 1960s as a positive-sum game. Popular sectors were not part of this consensus. In Argentina, critical partners to a consensus—both in civil society and within the state—were too fractious to conceive of anything else but zero-sum arrangements in their own benefit. The "zero-sum society" characterization of Argentina seems quite apt here (Calvert and Calvert 1989). See O'Donnell (1976), Kaufman (1979), Potash (1980), Sábato and Schvarzer (1983), Rouquié (1987), W. Smith (1991), and Erro (1993).

7. CNEA monopolized nuclear decisions thanks to the tripartite division of industrial sectors among the three armed services that gave CNEA, as the Navy's exclusive fiefdom, broad lateral autonomy.

8. Similar arrangements of state and private interests were prevalent among "energy subsystems" in the U.S. (Chubb 1983). Incidentally, CNEA's strong commitment to nuclear power despite Argentina's near self-sufficiency in oil preceded the 1973 energy crisis.

9. Haggard (1990:45).

10. Moore (1983). Calder (1988) discusses the impact of overlapping bureaucratic jurisdictions on Japan's foreign policy in high-tech areas. Policy in such cases is erratic and reactive. On the consequences of a highly segmented policy process in the French oil sector, see Feigenbaum (1985).

11. On the scientific and administrative autonomy of France's nuclear program, see Scheinman (1965:12), Gilpin (1968:165), Nau (1974), Wilsford

(1989:161), and Jasper (1990). In France, neither the parliament nor other public groups were allowed to influence nuclear policy in a meaningful way (Zysman 1983). On Japan's nuclear sector, see Samuels (1987) and Low (1994). On India's nuclear industry, see Mirchandani and Namboodiri (1981), Surrey (1987), and Kapur (1994).

12. On the origins and outcome of lateral autonomy in the PRC's arms industry, see J. Lewis et al. (1991).

13. On how a more coherent industrializing strategy can moderate the leverage of an autonomous agency, see Encarnation and Wells (1985).

14. As argued, "more finely tuned analyses must probe actual state organizations in relation to one another, in relation to past policy initiatives, and in relation to the domestic and transnational contexts of state activity" (Evans et al. 1985). On the limits of "institutional drive" as an explanatory variable, see Gereffi (1978), Johnson (1982), Kitschelt (1982), Grieco (1984), and Barzelay (1986). Samuels (1987), Friedman (1988), and Okimoto (1989) warn against looking exclusively at Japan's economic bureaucracy in understanding state intervention.

15. On the role of institutions in industrial policy, see, inter alia, Katzenstein (1978), Johnson (1982), and Hart (1992).

16. On the relevance of domestic political structures to developmental strategies and outcomes, see Ruggie (1983).

17. Aspirations for a leading political role do not preclude continued reliance on state-controlled political and economic change (Cardoso 1986).

18. On market versus strategic uncertainty, see Barzelay (1986).

19. Kuhn (1966). In France, the United Kingdom, and Canada, the state and nationalized utilities led the development of the nuclear industry, whereas in the United States and West Germany, private entrepreneurs became, after initial stimulation, the engine of the industry's growth. The U.S. government invested twice as much as private firms in reactor R&D and provided an array of subsidies (Hertsgaard 1983).

20. Surrey and Walker (1981); Kitschelt (1991).

21. On the extant literature's focus on the United States, European, and Japanese cases, see inter alia, DeLeon (1979), Kitschelt (1982), Jacobsen and Hofhansel (1984), Campbell (1988), Samuels (1987), and Jasper (1990). Kitschelt (1982) emphasizes structural conditions and the institutionalization of policies in organizations as major explanations of state intervention in nuclear industries and regards bureaucratic institutions and the structural relations of power expressed in their organization as providing partial explanation for policy differences. Campbell (1988) finds the nature of public-private links critical in explaining nuclear power decline in the United States.

22. For an excellent overview of the neoliberal consensus in Latin America, see W. Smith et al. (1994).

23. On win-sets, involuntary defection, and uncertainty in bargaining, see Putnam (1988).

24. On the need to temper the benefits of a centralized body entrusted with bargaining with foreign firms, see Encarnation and Wells (1985).

25. Gereffi (1983).

26. Low consensus within the PLO is evident not only regarding macro-political objectives in peacemaking (with Hamas representing a powerful challenger) but in the definition of economic futures as well. At the same time, PLO objectives and implementation still appear to be the product of a highly centralized process with Chairman Yasir Arafat at its center. The authority created to handle economic transfers has enjoyed little autonomy, leading to continued threats by the head of this agency to resign from the position.

27. On automobiles in Mexico, see Bennett and Sharpe (1985). On computers in India, see Grieco (1984). On how segmented (or, in their own term, "diffused") decision making increases the unpredictability of bargaining outcomes (and the costs to the investor), see Encarnation and Wells (1985).

28. On "structural power," see Caporaso (1978). I use it here to point out that levels of consensus and autonomy underwrite the rules shaping the bargaining context: who has the right to bargain, over what scope of issues (local content, pricing, technology transfer), and with what degree of independence.

29. On weak and strong states, see Zysman (1983), Ikenberry (1988), Putnam (1988), and Migdal (1988).

30. The mere possibility of establishing nuclear industries points to the limits of a deterministic approach that zeroes in on constraints imposed by the international system. The (not insignificant) barriers to entry in the manufacture of reactors (large capital investments, complexity of the technology, economies of scale, skilled personnel) can be offset by the growth of a local capacity in capital goods, engineering, and a managerial infrastructure.

31. The relative bargaining strength of the recipient may not necessarily decrease in proportion to the sophistication of the technology but to the latter's rate of change (Baranson and Roark 1985).

32. This aspect of Brazil's behavior also suggests that a privileged bargaining position (potential) is not always translated into an optimal—or even an effective—(actual) bargaining outcome.

33. On India's strong reluctance to rely on suppliers' credits, see Encarnation (1989).

34. On the importance of understanding sectoral variations, see Kitschelt (1991).

35. Bunker (1986).

36. For a comprehensive study of Brazil's state enterprises between 1967 and 1979, see Trebat (1983). There were 422 state enterprises in Brazil in 1985 (*Fatos*, May 6, 1985).

37. Encarnation (1989) explains how widespread this model was in Brazil. State firms used their privileged access to capital markets in order to expand into areas claimed by the private sector (Baer 1983). As Barzelay (1986) ar-

gued, Petrobrás had come to represent the interests and ideology of state capitalism.

38. Trebat (1983). Petrobrás created its own transport and construction subsidiaries that competed with established private firms.

39. Eletrobrás accounted for $12 billion of Brazil's total foreign debt in 1985 (*O Estado de São Paulo*, Oct. 22, 1985, p. 41; *O Globo*, Apr. 7, 1985; *Relatório* 6, no. 6, pp. 11, 16). On the power plant industry, see Newfarmer (1985) and Rosa (1978). Brazilian entrepreneurs had challenged Eletrobrás's preference for foreign equipment since the 1950s (*Relatório* 6, no. 5, p. 285).

40. *Brasil Energia*, Apr. 1982, p. 13.

41. Trebat (1983). On Petrobrás's arm's-length attitude toward national firms, see Villela (1976:79).

42. CVRD also created its own transport and construction subsidiaries, displacing private firms from that market (Barros 1978:302). CSN became the largest steel engineering firm in Brazil (Villela 1976). On Siderbrás, CVRD, and the private sector battle against state expansion in steel, see B. Schneider (1991).

43. Behrman (1980). On foreign licensing, see Erber (1977:177). Petrobrás's subsidiary in the petrochemical sector (Petroquisa Petrobrás Química) relied heavily on the importation of basic engineering for its refineries.

44. Brazilian subsidiaries of multinationals supplied the main components (85 percent of total capital goods supplies) (Rosa 1978).

45. *Brasil Energy*, Aug. 10, 1980, and Sept. 24, 1980.

46. *Brasil Energy*, Sept. 24, 1983.

47. Brazil's President Ernesto Geisel and his Minister of Mines and Energy Shigeaki Ueki helped shape both Petrobrás and Nuclebrás.

48. On CEME, see Gereffi (1983:229–31) and Evans (1976:133–36).

49. The term *mavericks* in the heading, as in Chapter 3, applies to state agencies taking unorthodox, idiosyncratic, and at times even recalcitrant, stands; it does not imply invariably successful or desirable industrial programs.

50. The greatest challenges to CNEA came, in fact, from Army-dominated military regimes in 1967 (Onganía), 1972 (Lanusse), and 1976 (Videla), who favored a nuclear program similar to Brazil's. The Army was far more cost-sensitive or efficiency-oriented with respect to CNEA's investments than to its own. On the declining power of the Navy and interservice struggles in Argentina, see Potash (1980).

51. Many of these sectors were more attractive to private investment than nuclear components because of lower market uncertainties. On the displacement of private firms by Army-controlled enterprises, see Poneman (1987:100–103). On President Arturo Frondizi's efforts to privatize Army-controlled industries, see Most (1991).

52. The Army concentrated on ferrous metals and chemicals, whereas the Air Force controlled nonferrous metals, such as aluminum. Private investors had no more than 1 percent equity at SOMISA, the state-owned steel firm.

About 80 percent of DGFM's output in the 1960s was for civilian use by major Army-controlled enterprises (YPF, Gas del Estado, Ferrocarriles Argentinos) and private industry (Rouquié 1982: 321–22). YPF was formally a tripartite protectorate of the three services. DGFM accounted for up to 5 percent of the country's GNP (Poneman 1987: 101), swallowed over 7 percent of the national budget (P. Lewis 1990:451), and accumulated over $1.5 billion in foreign debt (Varas 1989:53).

53. On the statist tradition of the Army, its aversion to privatization, and its consistent preference for foreign equipment and technology, see Mallon and Sorrouille (1975) and P. Lewis (1990).

54. Millán (1986:37).

55. A solid group of researchers at the Army's R&D center (CITESA) never enjoyed access to the inherently unstable power centers within the Army and the military industries (DGFM) and was largely disconnected from industry. This group was thus unable to rely on a stable protective political network, such as that created by CNEA.

56. Neither can strategic-military considerations explain differences between the Brazilian and Argentine nuclear industrial programs, even though both countries were ruled by the military for most of the relevant periods of nuclear development. Ironically, civilian regimes (Illía in Argentina, Quadros and Goulart in Brazil) were more strongly committed to mercantilistic objectives, or greater national entrepreneurial and technical participation in their nuclear industries, than many among the military.

57. Rouquié (1982); Conca (1992).

58. On the highly segmented nature of decision making in this area at the time, see Ramamurti (1987:213, 230). On the similarity in macropolitical and institutional conditions affecting both the nuclear and computer sectors in the early 1960s, see Langer (1989:98–100). CAPRE was an interministerial commission created in 1972 accountable to the Ministry of Planning; it was abolished in 1979, with the creation of SEI (Cline 1987:35–37).

59. The Navy's institutional interest in computers related to its efforts to expand a fleet that required computerized navigation and firing gear (Ramamurti 1987:220). See also Langer (1989:101–6). Here again, as in Argentina, the British-oriented Navy embraced state subsidiarity, cajoling private firms to undertake the task; it also co-opted national technical-scientific clienteles from a relatively large Brazilian pool.

60. Some preliminary decisions to exclude multinational manufacturing of minicomputers were taken in 1977–78 (Grieco 1984). About 270 Brazilian companies controlled 48 percent of the microcomputer (hardware and software) market by 1982, whereas foreign producers not affected by the law supplied mainframe computers (Spero 1982). On the policy of "market reserve" in informatics, see Erber (1985), Frischtak (1986), Evans (1986), Westman (1985), and Adler (1987).

61. By 1981 private sector representatives were formally part of SEI's Advisory Council.

62. Ramamurti (1987:175–76); Conca (1992:114–23). The only two private firms in the sector—NEIVA and AEROTEC—became EMBRAER's subconstractors and suppliers. For a pioneer study of EMBRAER, see Franko-Jones (1986).

63. By 1983, 223,000 small- and medium-sized Brazilian companies owned over 90 percent of EMBRAER's nonvoting stock, whereas the Ministry of Aeronautics controlled 55 percent of its voting stock (Ramamurti 1987:182–94).

64. Dahlman (1984).

65. About 350 enterprises (including 300 subsuppliers) employed over 200,000 workers (Chudnovsky and Nagao 1983:99; Brigagão 1984).

66. On Brazil's arms industry, see Dagnino (1983), Brigagão (1984), Tollefson (1991), Franko-Jones (1992), and Conca (1992).

67. The term is Hirschman's (1971).

68. Westphal et al. (1985:170). This paucity is being replaced with a growing literature, particularly since the late 1980s, on technological change across countries, sectors, and firms.

69. Maxwell (1981). On the conceptual and policy-related limitations of economic analysis in the study of technological change, see Clark (1985: 229–35). On the impact of the institutional context on technological trajectories, see Ergas (1987).

70. J. M. Katz (1980) uses the term *pari passu* to convey the almost inevitability of gradual accumulation of learning processes. Sercovich (1981:137) refers to "by-product" learning as "the standard learning by doing acquired in an almost compulsive way, not at all by intent, whenever a productive activity is performed."

71. On technology and learning, see Dore (1984) and Bell (1984). On the debate over whether the technological gap between developing and technologically advanced countries is widening, stagnant, or narrowing, see Fransman and King (1984), Landes (1992), and Amsden (1992).

72. Rosenberg (1982).

73. Hirschman (1967).

74. Cohen and Zysman (1987); Burton (1993). For other views on the relationship between manufacturing, technological advances, and national competitiveness, see Guile and Brooks (1987), Richardson (1990), and L. Cohen and Noll (1991).

75. Tyson (1993:12).

76. On the Clinton administration's industrial technology policy, see Burton (1993). The business community in the U.S. began courting greater governmental support for R&D activities to maintain international competitiveness. Of course, defense-related industries had long been the recipients of such support—through procurement policies in particular—throughout the Cold War era.

77. India, South Korea, Pakistan, Mexico, and Taiwan have operating nuclear power plants. Cuba, Iran, and the Philippines have plants at different

stages of construction. Prospective newcomers include Turkey, Egypt, Algeria, Israel, Indonesia, Malaysia, Thailand, and Nigeria, among others (Hu and Woite 1993:2–4). The People's Republic of China projects the expansion of its single operating power plant into a large-scale nuclear industry.

78. I addressed some of the motivations for pursuing nuclear energy only when they related to choices of industrial structure and technology. That discussion made clear that respective levels of energy sufficiency per se cannot explain those choices.

79. West Germany, France, Great Britain, and Japan all adopted the basic U.S. design of pressurized water reactors.

80. On the analytical weakness of rhetorical categories such as "autonomous," "self-reliant," and "dependent" policies, and the utility of studying actions, processes, and outcomes, see Katzenstein (1978) and Tooze (1984).

81. For a dedicated treatment of the domestic political sources of nuclear restraint (that is, of establishing nuclear-weapons-free zones) in different regions, see Solingen (1994b).

82. On urban, environmental, regional, and social consequences of Brazil's nuclear program, see Rosa and Hesles (1984).

83. Reppy (1985); Lichtenberg (1985); Rosenberg (1986); Gummett and Reppy (1988).

84. Burton (1993).

85. On "heartland" technologies, see Kaplinsky (1984:141).

86. Reppy (1985). In light of India's classic (and quite forceful) defense of the multiplier effects of nuclear programs, it might be useful to conduct a similar probe of India's record in this area.

87. Nuclear programs accounted for over 4 percent of each of these two countries' foreign debt throughout the 1980s.

References Cited

Abranches, Sergio H. de. 1978. "The Divided Leviathan: State and Economic Policy Formation in Authoritarian Brazil." Ph.D. diss., Cornell University.

Adler, E. 1987. *The Power of Ideology: The Quest for Technological Autonomy in Argentina and Brazil.* Berkeley: University of California Press.

Allison, Graham T. 1971. *Essence of Decision.* Boston: Little, Brown.

Almeida, F. L. de. 1983. *A expansão da industria de bens de capital: fatores determinantes.* Rio de Janeiro: Fundação Getulio Vargas.

Alonso, Marcelo. 1977. "Desarrollo tecnológico: conceptos y acciones." *Interciencia* 2, no. 5:288–91.

Alves, Maria Helena Moreira. 1983. "Mechanisms of Social Control of the Military Governments in Brazil 1964–1980." In A. R. M. Ritter and D. H. Pollock, eds., *Latin American Prospects for the 1980s.* New York: Praeger Special Studies.

Alves, R. N., et al. 1977. "Requirements for and Development of Trained Manpower Resources." Paper presented at the International Conference on Nuclear Power and Its Fuel Cycle, Salzburg, Austria.

Ames, Barry. 1987. *Political Survival: Politicians and Public Policy in Latin America.* Los Angeles: University of California Press.

Amsden, Alice. 1992. "Hyperbolizing Knowledge Is a Dangerous Thing: Reply to David Landes." *Contention* 1, no. 3 (Spring): 109–32.

Anglade, Christian. 1985. "The State and Capital Accumulation in Contemporary Brazil." In C. Anglade and C. Fortin, eds., *The State and Capital Accumulation in Latin America*, pp. 52–138. London: Macmillan Press.

Aráoz, A., and C. Martínez-Vidal. 1974. *Ciencia e industria: Un caso argentino.* Organization of American States, Estudios sobre el Desarrollo Científico y Tecnológico, Washington, D.C.

Aráoz, A., and Francisco Sercovich. 1977. "Oferta de tecnología comercializable." Buenos Aires: CISEA, Serie Ciencia y Tecnologia 4.

Arian, Asher. 1989. *Politics in Israel: The Second Generation.* Rev. ed. Chatham, N.J.: Chatham House Publishers.

Ariga, Michiko. 1981. "Restrictive Business Practices and International Controls on Transfer of Technology." In Sagafi-Nejad, et al., eds., pp. 177–200.

Arrow, Kenneth. 1962. "The Economic Implications of Learning by Doing." *Review of Economic Studies* (June).

Avineri, Shlomo. 1972. *Hegel's Theory of the Modern State.* New York: Cambridge University Press.

Backhaus, Kurt, and Otto Wildgruber. 1982. "Status of the Brazilian-German Nuclear Cooperation Program." *Nuclear Europe* 5:15–22.

Baer, Werner. 1983. *The Brazilian Economy: Growth and Development.* New York: Praeger.

Baer, Werner, Richard Newfarmer, and T. Trebat. 1976. "On State Capitalism in Brazil: Some New Issues and Questions." *Inter-American Economic Affairs* 30, no. 3:69–91.

Baez, J. N., L. Pello, H. Espejo. n.d. "Certificación de personal dedicado a ensayos no destructivos: una herramienta responsable." Buenos Aires: Comisión Nacional de Energía Atómica.

Banco Interamericano de Desarrollo/Instituto para la Integración de America Latina (BID/INTAL). 1983. *La cooperación empresarial argentino-brasileña.* Buenos Aires.

Banco Nacional de Desenvolvimento Econômico e Social (BNDES). 1981. *Relatório de atividades.* Rio de Janeiro: Sistema BNDES, Ministério da Indústria e do Comercio.

———. 1982a. *Annual Report.* Rio de Janeiro: Sistema BNDES, Ministério da Indústria e do Comercio.

———. 1982b. *30 años de BNDES: avaliação e rumos.* Rio de Janeiro: Sistema BNDES, Ministério da Indústria e do Comercio.

———. 1983. *Annual Report.* Rio de Janeiro: Sistema BNDES, Ministério da Indústria e do Comercio.

———. 1984. *Annual Report.* Rio de Janeiro: Sistema BNDES, Ministério da Indústria e do Comércio.

Baranson, Jack. 1978. *Technology and the Multinationals: Corporate Strategies in a Changing World Economy.* Lexington, Mass.: Lexington Books.

Baranson, Jack, and R. Roark. 1985. "Trends in North-South Transfer of High Technology." In Nathan Rosenberg and Carlos Frischtak, eds., *International Technology Transfer: Concepts, Measures, and Comparisons,* pp. 24–43. New York: Praeger.

Barber, Bernard. 1952. *Science and the Social Order.* Glencoe, Ill.: Free Press.

Barbosa, A. L. F. 1981. *Propiedade e quase-propriedade no comércio de tecnologia.* Brasília: Conselho Nacional de Pesquisas.

Barros, Alexandre de S. C. 1978. "The Brazilian Military: Professional Socialization, Political Performance and State Building." Ph.D. diss., University of Chicago.

———. 1984. "The Formulation and Implementation of Brazilian Foreign Policy: Itamaraty and the New Actors." In Heraldo Muñoz and Joseph S. Tulchin, eds., *Latin American Nations in World Politics*, pp. 30–44. Boulder, Colo.: Westview Press.

Barros, J. R. Mendonça de, and Douglas H. Graham. 1978. "The Brazilian Economic Military Revisited: Private and Public Sector Initiative in a Market Economy." *Latin American Research Review* 13, no. 2:5–38.

Bartolini, Stefano. 1993. "On Time and Comparative Research." *Journal of Theoretical Politics* 5, no. 2 (Apr.): 131–68.

Barzelay, Michael. 1986. *The Politicized Market Economy*. Berkeley: University of California Press.

Bates, Robert. 1981. *Markets and States in Tropical Africa: The Political Basis of Agricultural Policies*. Berkeley: University of California Press.

Bath, C. Richard, and Dilmus D. James. 1976. "Dependency Analysis of Latin America." *Latin American Research Review* 11, no. 3:3–54.

———. 1979. "The Extent of Technological Dependence in Latin America." In James H. Street and Dilmus D. James, eds., *Technological Progress in Latin America: The Prospects for Overcoming Dependency*. Boulder, Colo.: Westview Press.

Baumgartner, Frank R., and David Wilsford. 1994. "France: Science within the State." In Etel Solingen, ed., pp. 119–61.

Becker, David G. 1984. "Development, Democracy and Dependency in Latin America: a Postimperialist View." *Third World Quarterly* 6, no. 2:411–31.

———. 1990. "Business Associations in Latin America: The Venezuelan Case." *Comparative Political Studies* 23, no. 1 (Apr.): 114–38.

Becker, David G., J. Frieden, S. P. Schatz, and R. L. Sklar. 1987. *Postimperialism—International Capitalism and Development in the Late Twentieth Century*. Boulder, Colo.: Lynne Rienner.

Behrman, J. N. 1980. *Industry Ties with Science and Technology: Policies in Developing Countries*. Cambridge, Mass.: Oelgeschlager, Gunn and Hain.

Bell, M., D. S. Kemmis, and W. Satyarakwit. 1982. "Limited Learning in Infant Industry: A Case-study." In Frances Stewart and J. James, eds., pp. 138–56.

Bell, R. M. N. 1984. "'Learning' and the Accumulation of Industrial Technological Capacity in Developing Countries." In M. Fransman and K. King, eds., *Technological Capability in the Third World*, pp. 187–210. London: Macmillan.

Ben-David, Joseph. 1972. "The Profession of Science and Its Powers." *Minerva* 10:362–83.

Bendor, Jonathan, and T. H. Hammond. 1992. "Rethinking Allison's Models." *American Political Science Review* 86 (Winter): 301–22.

Bennett, Douglas C., and Kenneth E. Sharpe. 1979. "Agenda Setting and Bargaining Power: The Mexican State versus Transnational Automobile Corporations." *World Politics* 32, no. 1 (Oct.): 57–89.

———. 1985. *Transnational Corporations versus the State.* Princeton: Princeton University Press.

Biasi, Renato de. 1979. *A energia nuclear no Brasil.* Rio de Janeiro: Atlântida.

Biato, F., E. Guimarães, and M. H. Figuereido. 1971. "Potencial de pesquisa tecnológica no Brasil." Brasília: Instituto de Planejamento Econômico e Social (IPEA)/Instituto de Planejamento (IPLAN), Relatório de Pesquisas no. 5.

———. 1973. "A transferência de tecnologia no Brasil." Brasília: Instituto de Planejamento Econômico e Social (IPEA)/Instituto de Planejamento (IPLAN), Estudos para o planejamento no. 4.

Bond, J. S., and N. Rao. 1976. *Governmental Organization and Planning for Science and Technology in Brazil.* MIT: Center for Policy Alternatives.

Borrus, Michael, and John Zysman. 1986. "Japan." In F. W. Rushing and C. G. Brown, eds., *National Policies for Developing High Technology Industries: International Comparisons.* Boulder, Colo.: Westview Press.

Boschi, Renato R. 1979. *Elites industriais e democracia (hegemonia burguesa e mudança política no Brasil).* Rio de Janeiro: Edições GRAAL.

Braybrooke, David, and Charles E. Lindblom. 1963. *A Strategy of Decision.* New York: Free Press.

Bresser Pereira, Luiz C. 1978. *O colapso de uma aliança de classes: a burguesia e a crise do autoritarismo tecno-burocrático.* São Paulo: Editora Brasiliense.

———. 1981. *Estado e subdesenvolvimento industrializado.* São Paulo: Brasiliense.

———. 1984. "Six Interpretations of the Brazilian Social Formation." *Latin American Perspectives* 40, 11, 1:35–72.

———. 1988. "Possible Political Pacts After Redemocratization." In J. M. Chacel, P. S. Falk, and D. Fleischer, eds., *Brazil's Economic and Political Future*, pp. 141–52. Boulder, Colo.: Westview Press.

Brigagão, Clovis. 1984. *O mercado da Segurança.* Rio de Janeiro: Nova Fronteira.

Brito, Sergio de Salvo. 1967. "The Brazilian Advanced Thorium Converter." In Craig B. Smith, ed., *Nuclear Energy and Latin American Development*, pp. 215–18. Nuclear Energy Laboratory, University of California, Los Angeles.

Bruneau, Thomas C., and Phillipe Faucher, eds. 1981. *Authoritarian Capitalism: Brazil's Contemporary Economic and Political Development.* Boulder, Colo.: Westview Press.

Bunker, S. G. 1986. "Debt and Democratization: Changing Perspectives on the Brazilian State." *Latin American Research Review* 21, no. 3:206–23.

Burton, Daniel F., Jr. 1993. "High-Tech Competitiveness." *Foreign Policy* 92 (Fall): 117–32.

Calder, Kent. 1988. "Japanese Foreign Economic Policy Formation: Explaining the Reactive State." *World Politics* 40 (July): 517–41.

Calvert, Susan, and Peter Calvert. 1989. *Argentina: Political Culture and Instability.* Pittsburgh: University of Pittsburgh Press.

Campbell, John L. 1988. *Collapse of an Industry: Nuclear Power and the Contradictions of U.S. Policy.* Ithaca: Cornell University Press.

Canak, William L. 1984. "The Peripheral State Debate: State Capitalist and Bureaucratic-Authoritarian Regimes in Latin America." *Latin American Research Review* 19, no. 1 : 3–36.

Caporaso, James A. 1978. "Dependence, Dependency, and Power in the Global System: A Structural and Behavioral Analysis." *International Organization* 32, no. 1 (Winter): 13–43.

———, ed. 1989. *The Elusive State: International and Comparative Perspectives.* Newbury Park, Calif.: Sage.

Caputo, Dante. 1978. "State Purchasing Power as an Instrument for Technological Development." Buenos Aires: Centro de Investigaciones Sociales sobre el Estado y la Administración.

Cardoso, Fernando H. 1973. "Associated-Dependent Development: Theoretical and Practical Implications." In Alfred Stepan, ed., *Authoritarian Brazil*, pp. 142–78.

———. 1975. *Autoritarismo e Democratização.* Rio de Janeiro: Editora Paz e Terra.

———. 1978. *Política e desenvolvimento em sociedades dependentes: ideologias de empresariado industrial argentino e brasileiro.* Rio de Janeiro: Zahar.

———. 1986. "Entrepreneurs and the Transition Process: The Brazilian Case." In Guillermo O'Donnell, Philippe C. Schmitter, and Laurence Whitehead, eds., *Transitions from Authoritarian Rule—Comparative Perspectives*, pp. 137–53. Baltimore: Johns Hopkins University Press.

Cardoso, Fernando H., and E. Faletto. 1978. *Dependency and Development in Latin America.* Berkeley: University of California Press.

Carvalho, Joaquim F. 1981a. "Panorama Energético Brasileiro: O papel da nucleoelectricidade." In David Simon et al.

———. 1981b. "Aspectos económicos e estratégicos do acordo nuclear Brasil-Alemanha." *Ciencia e Sociedade.* Rio de Janeiro: Centro Brasileiro de Pesquisas Físicas.

———. 1985. "O que fazer com a NUCLEN." *O Estado de São Paulo*, 6 July.

Carvalho, Joaquim F., and José Goldemberg. 1980. *Economia e política da energia.* Rio de Janeiro: José Olympio.

Carvalho, Joaquim F., et al. 1987. *O Brasil nuclear—uma anatomia do desenvolvimento nuclear brasileiro.* Porto Alegre: Tche!.

Castro, A. Pereira de. 1974. "A organização de uma infraestrutura tecnologica para o desenvolvimento industrial brasileiro." *Revista de Administração de Empresas* 14, no. 3.

Cavarozzi, Marcelo. 1986. "Political Cycles in Argentina since 1955." In Guillermo O'Donnell, et al., eds., pp. 19–48.

Chang, Ha-Joon. 1994. *The Political Economy of Industrial Policy.* New York: St. Martin's Press.

Charpentier, Jean-Pierre, and Leonard Bennett. 1985. *IAEA Bulletin* 27, no. 4 (Winter): 41–46.

Chubb, John E. 1983. *Interest Groups and the Bureaucracy— The Politics of Energy.* Stanford: Stanford University Press.

Chudnovsky, D., and M. Nagao. 1983. *Capital Goods Production in the Third World.* New York: St. Martin's Press.

Cibotti, Ricardo, and Jorge Lucángeli. 1980. *El fenómeno tecnológico interno.* Buenos Aires: Banco Interamericano de Desarrollo/Comisión Económica para América Latina (CEPAL). Monografía de trabajo no. 29.

Clark, Norman. 1985. *The Political Economy of Science and Technology.* New York: Basil Blackwell.

———. 1987. "Similarities and Differences Between Scientific and Technological Paradigms." *Futures* (Feb.): 26–42.

Cline, William R. 1987. *Informatics and Development: Trade and Industrial Policy in Argentina, Brazil, and Mexico.* Washington, D.C.: Economics International.

Coelho, Edmundo Campos. 1976. *Em busca de identidade: o exército e a política na sociedade brasileira.* Rio de Janeiro: Editora Forense–Universitaria Ltda.

Cohen, Linda R., and Roger G. Noll. 1991. *The Technology Pork Barrel.* Washington, D.C.: Brookings Institution.

Cohen, Stephen S., and John Zysman. 1987. *Manufacturing Matters.* New York: Basic Books.

Cohen, Youssef. 1989. *The Manipulation of Consent: The State and Working-Class Consciousness in Brazil.* Pittsburgh, Pa.: University of Pittsburgh Press.

Collier, David, ed. 1979. *The New Authoritarianism in Latin America.* Princeton: Princeton University Press.

Collier, Ruth B., and David Collier. 1991. *Shaping the Political Arena: Critical Junctures, the Labor Movement, and Regime Dynamics in Latin America.* Princeton: Princeton University Press.

Comisión Nacional de Energía Atómica (CNEA). 1981. *Memoria anual.* Buenos Aires.

———. 1982. *Memoria anual.* Buenos Aires.

———. 1983. *Memoria anual.* Buenos Aires.

———. n.d. "La dirección de investigación y desarrollo y sus programas." Buenos Aires.

Comissão Nacional de Energia Nuclear (CNEN). 1980. *Relatório anual.* Rio de Janeiro.

Companhia Brasileira de Tecnologia Nuclear (CBTN). 1973. "Capacidade da industria brasileira para fabricação de equipamentos de usinas nucleares." Rio de Janeiro: Bechtel.

Conca, Kenneth L. 1992. "Global Markets, Local Politics, and Military Industrialization in Brazil." Ph.D. diss., University of California, Berkeley.

Cooper, Charles. 1972. "Science, Technology and Production in the Underdeveloped Countries: An Introduction." *Journal of Development Studies* 9, no. 1:1–18.

———. 1973. "Choice of techniques and technological change as problems in political economy." *International Social Science Journal* 25, no. 3:293–304.

———. 1974. "Science Policy and Technological Change in Underdeveloped Economies." *World Development* 2, no. 3:55–64.

Cooper, C. M., and F. Sercovich. 1971. "The Channels and Mechanisms for the Transfer of Technology from Developed to Developing Countries." Geneva: UNCTAD, TD/B/AC.11/5, 27 April.

Cortés, Mariluz, and Peter Bocock. 1984. *North-South Technology Transfer—A Case Study of Petrochemicals in Latin America*. Baltimore: Johns Hopkins University Press.

Cosentino, Jorge. 1984. "La experiencia de desagregación en el programa de reactores y centrales nucleares argentino." Paper presented at the International Seminar on Project Disaggregation by Public Enterprises in Developing Countries, Buenos Aires.

Costa, J. Ribeiro da. 1982. "Quality Assurance for Nuclear Power Plants." Vienna: International Atomic Energy Agency.

Courtney, William H. 1980. "Nuclear Choices for Friendly Rivals." In Joseph A. Yager, ed., *Nonproliferation and U.S. Foreign Policy*, pp. 241–79. Washington, D.C.: Brookings Institution.

Courtney, W. H., and D. M. Leipziger. 1974. "MNCs in LDCs: The Choice of Technology." Washington, D.C.: AID, Discussion Paper no. 29.

Cox, Robert W. 1986. "Social Forces, States and World Orders: Beyond International Relations Theory." In Robert O. Keohane, ed., *Neorealism and Its Critics*, pp. 204–54. New York: Columbia University Press.

———. 1987. *Production, Power, and World Order: Social Forces in the Making of History*. Vol. 1. New York: Columbia University Press.

Cyert, R. M., and J. G. March. 1963. *A Behavioral Theory of the Firm*. Englewood Cliffs, N.J.: Prentice-Hall.

Dagnino, Renato P. 1983. "Indústria de armamentos: o estado e a tecnologia." *Revista Brasileira de Tecnologia* 14, no. 3:5–17.

Dahl, Robert. 1961. *Who Governs: Democracy and Power in an American City*. New Haven: Yale University Press.

Dahlman, Carl. 1984. "Foreign Technology and Indigenous Technological Capability in Brazil." In M. Fransman and K. King, eds., pp. 317–34.

Dahlman, Carl J., Bruce Ross-Larson, and Larry E. Westphal. 1985. "Managing Technological Development—Lessons from the Newly Industrializing Countries." World Bank Staff Working Papers, no. 717. Washington, D.C.: World Bank.

Dahlman, Carl, and Francisco C. Sercovich. 1984. "Local Development and Exports of Technology." World Bank Staff Working Papers, no. 667. Washington, D.C.: World Bank.

Dahlman, Carl, and Larry Westphal. 1982. "Technological effort in industrial development—an interpretative survey of recent research." In Frances Stewart and J. James, eds., *The Economics of New Technology in Developing Countries*. Boulder, Colo.: Westview Press.

David, Paul A. 1985. "Clio and the Economics of QWERTY." *Journal of Economic History* 75, no. 2 (May): 332–37.

Dedijer, Stevan. 1968. "Underdeveloped Science in Underdeveloped Countries." In Edward Shils, ed., *Criteria for Scientific Development: Public Policy and National Goals*. Cambridge, Mass.: MIT Press.

DeLeon, Peter. 1979. *Development and Diffusion of the Nuclear Power Reactor: A Comparative Analysis*. Cambridge, Mass.: Ballinger.

Deyo, Frederic C., ed. 1987. *The Political Economy of New Asian Industrialism*. Ithaca: Cornell University Press.

Díaz-Alejandro, Carlos F. 1970. *Essays on the Economic History of the Argentine Republic*. New Haven: Yale University Press.

Dickson, D. 1978. "Brazil Scientists Fan Doubts over Energy Priorities." *Nature* 275 (Oct. 19).

Diniz, Eli, and Renato R. Boschi. 1977. "Elite industrial e estado: uma análise da ideologia do empresariado nacional nos anos 70." In Carlos E. Martins, ed., *Estado e capitalismo no Brasil*, pp. 167–90. São Paulo: Humanismo, Ciencia e Tecnologia (HUCITEC), Centro Brasileiro de Análise e Planejamento (CEBRAP).

———. 1978. *Empresariado nacional e estado no Brasil*. Rio de Janeiro: Forense-Universitária.

Diniz, Eli, and O. Brasil de Lima, Jr. 1985. "Modernização autoritaria, centralização e intervenção do estado na economia." Rio de Janeiro: Instituto Universitário de Pesquisas do Rio de Janeiro (IUPERJ) (mimeo).

Di Tella, Guido, and Rudiger Dornbusch, eds. 1989. *The Political Economy of Argentina, 1946–83*. London: Macmillan.

Domínguez, Jorge I. 1982. *Economic Issues and Political Conflict: U.S.– Latin American Relations*. Boston: Butterworth Scientific.

———. 1987. "Order and Progress in Brazil." In George C. Lodge and Ezra F. Vogel, eds., *Ideology and National Competitiveness: An Analysis of Nine Countries*, pp. 241–70. Boston: Harvard Business School Press.

Doner, Richard F. 1992. "Limits of State Strength: Toward an Institutionalist View of Economic Development." *World Politics* 44, no. 3 (Apr.): 398–431.

Dore, Ronald. 1984. "Technological Self-Reliance: Sturdy Ideal or Self-Serving Rhetoric." In M. Fransman and K. King, eds., pp. 65–80.

Dorfman, Adolfo. 1983. *Cincuenta años de industrialización en la Argentina: 1930–1980*. Buenos Aires: Ediciones Solar.

Downs, Anthony. 1967. *Inside Bureaucracy*. Boston: Little, Brown.

Dreifuss, René A. 1981. *1964: a conquista do estado—ação política, poder e golpe de classe*. Petrópolis: Vozes.

Dreifuss, René A., and O. S. Dulci. 1983. "As forças armadas e a política." In Bernardo Sorj, et al., *Sociedade e política no Brasil pos-1964*, pp. 87–117. São Paulo: Brasiliense.

Dunning, John H. 1982. "Toward a Taxonomy of Technology Transfer and Possible Impacts on OECD Countries." In *North/South Technology Transfer — The Adjustments Ahead*, pp. 8–24. Paris: OECD.

Duvall, Raymond D., and John R. Freeman. 1981. "The State and Dependent Capitalism." In W. Ladd Hollist and James Rosenau, eds., pp. 223–42.

Dye, D. R., and C. E. de Souza e Silva. 1979. "A Perspective on the Brazilian State." *Latin American Research Review* 14, no. 1:81–98.

Eckstein, Harry. 1966. *Division and Cohesion in Democracy: A Study of Norway*. Princeton: Princeton University Press.

———. 1975. "Case Study and Theory in Political Science." In Fred Greenstein and Nelson Polsby, eds., *Handbook of Political Science*. Vol. 7, pp. 79–138. Reading, Mass.: Addison-Wesley.

El Desarrollo Nuclear Argentino. 1985. Buenos Aires: Edigraf.

Elkoubi, A., and J. Bernard. 1982. "Fuel Design and Manufacturing Technology Transfer: A Driving Force of Industrial Development." Paper presented at the Second International Conference on Nuclear Technology Transfer, Buenos Aires.

Elster, Jon. 1983. *Explaining Technical Change — A Case Study in the Philosophy of Science*. Cambridge, Eng.: Cambridge University Press.

Encarnation, Dennis J. 1989. *Dislodging Multinationals — India's Strategy in Comparative Perspective*. Ithaca: Cornell University Press.

Encarnation, Dennis J., and Louis T. Wells. 1985. "Sovereignty En Garde: Negotiating with Foreign Investors." *International Organization* 39 (Winter): 47–78.

Erber, Fabio. 1977. "Technological Development and State Intervention: A Study of the Brazilian Capital Goods Industry." Ph.D. diss., University of Sussex.

———. 1980. "Desenvolvimento tecnológico e intervenção do estado: um confronto entre a experiência brasileira e a dos países capitalistas centrais." *Revista de Administração Pública* 14, 4:10–72.

———. 1981a. "Science and Technology Policy: A View from the Periphery." In Joseph S. Szyliowicz, ed., *Technology and International Affairs*, pp. 173–200. New York: Praeger.

———. 1981b. "Science and Technology Policy in Brazil: A Review of the Literature." *Latin American Research Review*, no. 1:3–56.

———. 1984. "The Capital Goods Industry and the Dynamics of Economic Development in LDCs—The Case of Brazil." Rio de Janeiro, Universidade Federal do Rio de Janeiro, Instituto de Economia Industrial: Texto para discussão no. 48.

———. 1985. "Microeletrónica: Revolução ou reforma?" Instituto de Economia Industrial, Universidade Federal do Rio de Janeiro. Texto para discussão no. 6.

Ergas, Henry. 1987. "Does Technology Policy Matter?" In Bruce Guile and Harvey Brooks, eds., pp. 191–245.

Erickson, Kenneth P. 1977. *The Brazilian Corporative State and Working Class Politics*. Berkeley: University of California Press.

———. 1981a. "State Entrepreneurship, Energy Policy and the Political Order in Brazil." In Thomas Bruneau and Phillipe Faucher, eds., pp. 141–78.

———. 1981b. "The energy profile of Brazil." In Kenneth R. Stunkel, ed., *National Energy Profiles*. New York: Praeger.

Erro, Davide G. 1993. *Resolving the Argentine Paradox: Politics and Development 1966–1992*. Boulder, Colo.: Lynne Rienner.

Evans, Peter. 1976. "Foreign Investment and Industrial Transformation." *Journal of Development Economics* 3, no. 4:119–39.

———. 1979. *Dependent Development: The Alliance of Multinational, State, and Local Capital in Brazil*. Princeton: Princeton University Press.

———. 1982. "Reinventing the Bourgeoisie: State Entrepreneurship and Class Formation in Dependent Capitalist Development." In M. Burawoy and Theda Skocpol, eds., *Marxist Inquiries: Studies of Labor, Class, and States*. Chicago: University of Chicago Press.

———. 1985. "Transnational Linkages and the Economic Role of the State: An Analysis of Developing and Industrialized Nations in the Post–World War II Period." In Peter Evans, et al., eds., pp. 192–226.

———. 1986. "State, Capital, and the Transformation of Dependence: The Brazilian Computer Case." *World Development* 14:791–808.

Evans, Peter, Dietrich Rueschemeyer, and Theda Skocpol, eds. 1985. *Bringing the State Back In*. Cambridge, Eng.: Cambridge University Press.

Fabricio, R. A. C. 1982. "Brazil's General Experience in the Transfer of Nuclear Technology." Paper presented at the Second International Conference on Nuclear Technology Transfer, Buenos Aires.

Fagen, Richard R. 1977. "Studying Latin American Politics: Some Implications of a Dependencia Approach." *Latin American Research Review* 12, no. 2:3–26.

Fajnzylber, Fernando. 1971. "Sistema Industrial e Exportação de Manufaturados: Análise da Experiência Brasileira." Rio de Janeiro: IPEA Relatório de Pesquisas, No. 7.

Fajnsylber, F. 1977. "Oligopolio, empresas transnacionais e estilos de desenvolvimento." *Cadernos CEBRAP* no. 19:7–35.

Falicov, Leo M. 1970. "Physics and Politics in Latin America: A Personal Experience." *Bulletin of Atomic Scientists* 26, no. 11 (Nov.): 8–10, 41–43.

Faucher, Phillipe. 1980. "Industrial Policy in a Dependent State: The Case of Brazil." *Latin American Perspectives* 7, no. 1:3–22.

———. 1981. "The paradise that never was: the breakdown of the Brazilian authoritarian order." In Thomas Bruneau and Phillipe Faucher, eds., pp. 11–40.

Feigenbaum, Harvey B. 1985. *The Politics of Public Enterprise: Oil and the French State*. Princeton: Princeton University Press.

Felix, David. 1974. "Technological Dualism in Late Industrializers: On Theory, History, and Policy." *Journal of Economic History* 24:194–238.

Ferrer, Aldo. 1974. *Tecnología y política económica en América Latina*. Buenos Aires: Paidos.

———. 1981. "La economía argentina al comenzar la década de 1980." *El Trimestre Económico* 4, no. 192:809–51.

Fishlow, Albert. 1971. *American Railroads and the Transformation of the Ante-bellum Economy.* Cambridge, Mass.: Harvard University Press.

———. 1985. "The State of Latin American Economics." In *Economic and Social Progress in Latin America—External Debt: Crisis and Adjustment.* Washington, D.C.: InterAmerican Development Bank. 1985 Report: 123–45.

———. 1987. "Some Reflections on Comparative Latin American Economic Performance and Policy." Working Paper 8754. Department of Economics. University of California, Berkeley.

———. 1989. "A Tale of Two Presidents: The Political Economy of Crisis Management." In Alfred Stepan, *Democratizing Brazil,* pp. 83–119.

Flynn, Peter. 1978. *Brazil, A Political Analysis.* Boulder, Colo.: Westview Press.

Fontana, Andrés M. 1987. "Political decision making by a military corporation: Argentina 1976–1983." Ph.D. diss., University of Texas at Austin.

Ford, E., et al. 1977. "A oferta de serviço de consultoría de engenharia no Brasil." Rio de Janeiro: FINEP (mimeo).

Foreign Broadcast Information Service (FBIS). *Worldwide Report: Nuclear Development and Proliferation.* Washington, D.C.: National Technical Information Service.

Frame, D. J. 1983. *International Business and Global Technology.* Lexington, Mass.: Lexington Books.

Franko-Jones, Patrice. 1986. "The Brazilian Defense Industry: A Case Study of Public-Private Collaboration." Ph.D. diss., Notre Dame University.

———. 1992. *The Brazilian Defense Industry.* Boulder, Colo.: Westview Press.

Fransman, M. 1986. "International Competitiveness, Technical Change, and the State: The Machine Tool Industry in Taiwan and Japan." *World Development* 14, no. 12:1375–96.

Fransman, M., and K. King, eds., 1984. *Technological Capability in the Third World.* London: Macmillan.

Freeman, Christopher, C. J. Clark, and Luc Soete. 1982. *Unemployment and Technical Innovation: A Study of Long Waves and Economic Development.* London: Frances Pinter.

Frieden, Jeffry A. 1991. *Debt, Development and Democracy: Modern Political Economy and Latin America, 1965–1985.* Princeton: Princeton University Press.

Friedman, David. 1988. *The Misunderstood Miracle: Industrial Development and Political Change in Japan.* Ithaca: Cornell University Press.

Frischtak, Claudio. 1986. "Brazil." In Francis W. Rushing and Carole G. Brown, eds., *National Policies for Developing High Technology Industries: International Comparisons,* pp. 31–70. Boulder, Colo.: Westview Press.

Fuenzalida, Edmundo F. 1979. "The Problems of Technological Innovation in Latin America." In J. J. Villamil, ed., *Transnational Capitalism: New Perspectives on Dependence*, pp. 115–28. New Jersey: Humanities Press.

Fúnes, Amílcar. 1984. "Investigación, desarrollo y tecnología en el area nucleoenergética Argentina." Paper presented at the Panamerican Congress of Mechanical Engineers, Electricians, and Associated Specialties, Buenos Aires.

Furtado, Celso. 1976. *The Economic Development of Latin America*. Cambridge, Eng.: Cambridge University Press.

Galtung, Johan. 1978–79. "Towards a New International Technological Order." *Alternatives* 4:277–300.

Gasparian, A. E., and H. Calvet Filho. 1979. "The Role of Brazilian Architect Engineering Firms in the Transfer and Development of Nuclear Power Plant Technology." Paper presented at the Pan American Nuclear Technology Exchange Conference, Hollywood, Fla., Apr. 8–11.

Gattas, Ramiz. 1981. *A indústria automobilística e a 2a revolução industrial no Brasil—origens e perspectivas*. São Paulo: Prelo.

Geddes, Barbara. 1990. "Building 'State' Autonomy in Brazil, 1930–1964." *Comparative Politics* 22, no. 2 (Jan.): 217–36.

———. 1994. *Politician's Dilemma: Building State Capacity in Latin America*. Berkeley: University of California Press.

Gereffi, Gary. 1978. "Drug Firms and Dependency in Mexico: The Case of the Steroid Hormone Industry." *International Organization* 32, no. 1 (Winter): 237–86.

———. 1983. *The Pharmaceutical Industry and Dependency in the Third World*. Princeton: Princeton University Press.

———. 1990. "Big Business and the State." In Gary Gereffi and Donald Wyman, eds., pp. 90–109.

Gereffi, Gary, and Peter Evans. 1981. "TNCs, Dependent Development, and State Policy in the Semiperiphery: A Comparison of Brazil and Mexico." *Latin American Research Review* 16, no. 3:31–64.

Gereffi, Gary, and Donald L. Wyman, eds. 1990. *Manufacturing Miracles—Paths of Industrialization in Latin America and East Asia*. Princeton: Princeton University Press.

Gerschenkron, Alexander. 1962. *Economic Backwardness in Historical Perspective*. Cambridge, Mass.: Belknap Press.

Gilpin, Robert. 1962. *American Scientists and Nuclear Weapons Policy*. Princeton: Princeton University Press.

———. 1968. *France in the Age of the Scientific State*. Princeton: Princeton University Press.

———. 1987. *The Political Economy of International Relations*. Princeton: Princeton University Press.

Girotti, Carlos A. 1984. *Estado nuclear no Brasil*. São Paulo: Brasiliense.

Góes, Walder de. 1978. *O Brasil do General Geisel: Estudo do proceso de tomada de decisão no regime militar-burocrático*. Rio de Janeiro: Nova Fronteira.

Goldemberg, José. 1978. *Energia nuclear no Brasil—as origens das decisões.* São Paulo: Hucitec.

Goldsmith, Maurice, and Alexander King, eds. 1979. *Issues of Development: Towards a New Role for Science and Technology.* Oxford: Pergamon Press.

Goldwert, Marvin. 1972. *Democracy, Militarism, and Nationalism in Argentina, 1930–66.* Austin: University of Texas Press.

Gonzaga, L. A., and R. M. Hokama. 1982. "The Brazilian Experience in the Construction of Nuclear Power Plants." Paper presented at the Second International Conference on Nuclear Technology Transfer, Buenos Aires, Nov. 1–5.

Gonzaga, L. A., F. S. Marinho, and J. R. Batista. 1979. "Civil Construction Experience with Nuclear Power Plants in Brazil." Paper presented at the Pan American Nuclear Technology Exchange Conference, Hollywood, Fla., Apr. 8–11.

Goulet, Denis. 1977. *The Uncertain Promise: Value Conflicts in Technology Transfer.* New York: IDOC/North America.

Gourevitch, Peter. 1978. "The second image reversed: the international sources of domestic politics." *International Organization* 32:881–911.

———. 1986. *Politics in Hard Times—Comparative Responses to International Economic Crises.* Ithaca: Cornell University Press.

———. 1989. "Keynesian Politics: The Political Sources of Economic Policy." In Hall, ed., pp. 87–106.

Gramsci, Antonio. 1988. *A Gramsci Reader: Selected Writings 1916–1935.* Ed. David Forgacs. London: Lawrence and Wishart.

Grieco, Joseph M. 1984. *Between Dependency and Autonomy—India's Experience with the International Computer Industry.* Berkeley: University of California Press.

———. 1986. "Foreign Investment and Development: Theories and Evidence." In Moran, ed., pp. 35–52.

Guadagni, A. A. 1985. "La programación de las inversiones eléctricas y las actuales prioridades energéticas." *Desarrollo Económico* 25:98.

Guile, Bruce R., and Harvey Brooks, eds. 1987. *Technology and Global Industry—Companies and Nations in the World Economy.* Washington, D.C.: National Academy Press.

Guilherme, Olympio. 1957. *O Brasil e a era atômica.* Rio de Janeiro: Editorial Vitoria.

Gummett, Philip, and Judith Reppy, eds. 1988. *The Relations Between Defence and Civil Technologies.* Dordrecht, The Netherlands: Kluwer.

Haas, Ernst B. 1980. "Technological Self-Reliance for Latin America: the OAS contribution." *International Organization* 34, no. 4:541–70.

Haberer, Joseph. 1969. *Politics and the Community of Science.* New York: Van Nostrand Reinhold.

Haggard, Stephan. 1986. "The Newly Industrializing Countries in the International System." *World Politics* 38, no. 2 (Jan.): 343–70.

———. 1990. *Pathways from the Periphery — The Politics of Growth in the Newly Industrializing Countries*. Ithaca: Cornell University Press.

Haggard, Stephan, and Tun-jen Cheng. 1987. "State and Foreign Capital in East Asian NICs." In Frederic C. Deyo, ed., *The Political Economy of the New Asian Industrialism*, pp. 84–135. Ithaca, N.Y.: Cornell University Press.

Hall, Peter A. 1989. "Conclusions: The Politics of Keynesian Ideas." In Peter A. Hall, ed., pp. 361–91.

———, ed. 1989. *The Political Power of Economic Ideas: Keynesianism across Nations*. Princeton: Princeton University Press.

Halperin, Morton. 1974. *Bureaucratic Politics and Foreign Policy*. Washington, D.C.: Brookings Institution.

Hart, Jeffrey A. 1992. *Rival Capitalists — International Competitiveness in the United States, Japan, and Western Europe*. Ithaca: Cornell University Press.

Helleiner, G. K. 1975. "The Role of MNCs in the LDCs' Trade in Technology." *World Development* 3:161–89.

Herrera, Amílcar. 1972. "Social Determinants of Science Policy and Implicit Science Policy." *Journal of Development Studies* 9, no. 1:19–38.

Hertsgaard, Mark. 1983. *Nuclear Inc. — The Men and Money Behind Nuclear Energy*. New York: Pantheon Books.

Hewlett, Sylvia Ann. 1978. "The State and Brazilian Economic Development: The Contemporary Reality and Prospects for the Future." In William H. Overholt, ed., *The Future of Brazil*, pp. 149–209. Boulder, Colo.: Westview Press.

———. 1980. *The Cruel Dilemmas of Development: Twentieth-Century Brazil*. New York: Basic Books.

Hirschman, Albert O. 1958. *The Strategy of Economic Development*. New Haven: Yale University Press.

———. 1967. *Development Projects Observed*. Washington, D.C.: Brookings Institution.

———. 1970. *Exit, Voice, and Loyalty: Responses to Decline in Firms, Organizations, and States*. Cambridge, Mass.: Harvard University Press.

———. 1971. *A Bias for Hope*. New Haven: Yale University Press.

———. 1979. "The Turn to Authoritarianism in Latin America and the Search for its Economic Determinants." In David Collier, ed. (1979): 61–98.

———. 1981. *Essays in Trespassing: Economics to Politics and Beyond*. Cambridge, Eng.: Cambridge University Press.

———. 1982. *Shifting Involvements: Private Interest and Public Action*. Princeton: Princeton University Press.

———. 1986. *Rival Views of Market Society and Other Recent Essays*. New York: Viking.

Hodara, Joseph. 1983. "Hirschman y la dependencia: el eslabón olvidado." *Desarrollo Económico* 23, no. 90:299–305.

Holdren, John P. 1983. "Nuclear Power and Nuclear Weapons: the Connection is Dangerous." *Bulletin of Atomic Scientists* (Jan.): 40–45.

Hollis, Martin, and Steve Smith. 1990. *Explaining and Understanding International Relations*. Oxford: Oxford University Press.

Hollist, W. Ladd. 1981. "Brazilian Dependence: An Evolutionary, World System Perspective." In W. Ladd Hollist and James Rosenau, eds., pp. 202–22.

———. 1983. "Differential Growth and Peripheral Development: A Theory of Continuity and Change." Paper presented at the International Studies Association annual convention, Mexico D.F.

Hollist, W. Ladd, and James N. Rosenau, eds. 1981. *World System Structure — Continuity and Change*. Beverly Hills, Calif.: Sage.

Hollist, W. Ladd, and Lamond Tullis, eds. 1985. *An International Political Economy*. Boulder, Colo.: Westview Press.

Horowitz, Irving L., and Ellen K. Trimberger. 1976. "State Power and Military Nationalism in Latin America." *Comparative Politics* 8, no. 2:223–44.

Hu, Chuanwen, and Georg Woite. 1993. "Nuclear power development in Asia." *IAEA (International Atomic Energy Agency) Bulletin* 35, no. 4 (Dec.): 2–7.

Hveem, Helge. 1983. "Selective Dissociation in the Technology Sector." In John G. Ruggie, ed., *The Antinomies of Interdependence*, pp. 273–316. New York: Columbia University Press.

Ikenberry, John G. 1988. *Reasons of State — Oil Politics and the Capacities of American Government*. Ithaca: Cornell University Press.

Ikenberry, John G., David A. Lake, and Michael Mastanduno, eds. 1988. *The State and American Foreign Economic Policy*. Ithaca: Cornell University Press.

Imai, R., and H. S. Rowen. 1980. *Nuclear Energy and Nuclear Proliferation: Japanese and American Views*. Boulder, Colo.: Westview Press.

Imaz, José L. de. 1964. *Los que mandan*. Buenos Aires: Editorial Universitaria de Buenos Aires.

Instituto de Planejamento Econômico e Social (IPEA). 1984. *Engenharia e Consultoria no Brasil e no Grupo Andino: Possíveis Áreas de Cooperação*. Série Estudos para o Planejamento no. 25. IPEA: Brasília.

Jacobsen, John K. 1987. "Peripheral Postindustrialization, Ideology, High Technology, and Dependent Development." In James Caporaso, ed., *A Changing International Division of Labor*, pp. 91–122. Boulder, Colo.: Lynne Rienner.

Jacobsen, John K., and C. Hofhansel. 1984. "Safeguards and Profits: Civilian Nuclear Exports, Neo-Marxism, and the Statist Approach." *International Studies Quarterly* 28:195–218.

Jasper, James M. 1990. *Nuclear Politics: Energy and the State in the United States, Sweden and France*. Princeton: Princeton University Press.

Johnson, Chalmers. 1982. *MITI and the Japanese Miracle: The Growth of Industrial Policy, 1925–1975*. Stanford: Stanford University Press.

————. 1987. "Political Institutions and Economic Performance: The Government-Business Relationship in Japan, South Korea, and Taiwan." In Frederic Deyo, ed., pp. 136–64.

Kaplan, M. 1974. "Commentary on Ianni." In J. Cotler and R. Fagen, eds., *Latin America and the United States: The Changing Political Realities*. Stanford: Stanford University Press.

Kaplinsky, Raphael. 1984. "Trade in Technology: Who, What, Where, and When?" In M. Fransman and K. King, eds., pp. 139–60.

Kapur, Ashok. 1994. "India: The Nuclear Scientists and the State, the Nehru and Post-Nehru Years." In Etel Solingen, ed., pp. 209–30.

Katz, James E. 1982. "Scientists, Government and Nuclear Power." In James E. Katz and Onkar S. Marwah, eds., *Nuclear Power in Developing Countries*. Lexington, Mass.: Lexington Books.

Katz, Jorge M. 1976. *Importación de tecnología, aprendizaje, e industrialización dependiente*. Mexico D.F.: Fondo de Cultura Económica.

————. 1980. "Domestic Technology Generation in LDCs: A Review of Research Findings." Buenos Aires: United Nations Economic Commission for Latin America, Research Programme on Scientific and Technological Development in Latin America. Working Paper no. 35.

————. 1984. "Technological Innovation, Industrial Organization and Comparative Advantages of Latin American Metalworking Industries." In M. Fransman and K. King, eds., pp. 113–36.

————. 1985. "Domestic Technological Innovations and Dynamic Comparative Advantages: Further Reflections on a Comparative Case-Study Program." In Nathan Rosenberg and Carlos Frischtak, eds., *International Technology Transfer*, pp. 127–65. New York: Praeger Special Studies.

Katz, Jorge M., and Eduardo Ablin. 1978. "Technology and Industrial Exports: A Micro-Economic Analysis of Argentina's Recent Experience." Buenos Aires: InterAmerican Development Bank/Economic Commission for Latin America, Research Program in Science and Technology, Working Paper no. 2.

Katz, Jorge M., and Ricardo Cibotti. 1978. "Marco de referencia para un programa de investigación en temas de ciencia y tecnología en la América Latina." *El Trimestre Económico* 45, 1, 177:139–65.

Katzenstein, Peter J. 1978. "Conclusion: Domestic Structures and Strategies of Foreign Economic Policies." In Peter J. Katzenstein, ed., *Between Power and Plenty*, pp. 295–336. Madison: University of Wisconsin Press.

————. 1984. *Corporatism and Change: Austria, Switzerland, and the Politics of Industry*. Ithaca: Cornell University Press.

————. 1985. *Small States in World Markets: Industrial Policy in Europe*. Ithaca: Cornell University Press.

Kaufman, Robert R. 1979. "Industrial Change and Authoritarian Rule in Latin America: A Concrete Review of the Bureaucratic-Authoritarian Model." In David Collier, ed., pp. 165–254.

————. 1990. "How Societies Change Developmental Models or Keep Them:

Reflections on the Latin American Experience in the 1930s and the Postwar World." In Gary Gereffi and Donald Wyman, eds., pp. 110–38.

Kennedy, Paul. 1987. *The Rise and Fall of the Great Powers*. New York: Random House.

Keohane, Robert O., and Joseph S. Nye. 1977. *Power and Interdependence*. Boston: Little, Brown.

Keren, Michael. 1980. "Science versus Government: A Reconsideration." *Policy Sciences* 12:333–53.

King, Kenneth. 1984. "Science, Technology, and Education in the Development of Indigenous Technological Capability." In M. Fransman and K. King, eds., *Technological Capability in the Third World*, pp. 31–64. London: Macmillan.

Kitschelt, Herbert. 1982. "Structures and Sequences of Nuclear Energy Policy-Making: Suggestions for a Comparative Perspective." *Political Power and Social Theory* 3:271–308.

———. 1986. "Four Theories of Public Policy Making and Fast Breeder Reactor Development." *International Organization* 40, no. 1:65–104.

———. 1991. "Industrial governance structures, innovation strategies, and the case of Japan: sectoral or cross-national comparative analysis?" *International Organization* 45, no. 4 (Autumn): 453–94.

Kittl, Jorge E., and J. C. Almagro. 1982. "Manufacturing of Fuel Element Cladding Tubes." Paper presented at the Second International Conference on Nuclear Technology Transfer, Buenos Aires.

Klapp, Merrie Gilbert. 1987. *The Sovereign Entrepreneur: Oil Policies in Advanced and Less Developed Capitalist Countries*. Ithaca: Cornell University Press.

Knodler, D. 1982. "The Role of Quality Assurance and Specifications in Technology Transfer." Paper presented at the Second International Conference on Nuclear Technology Transfer, Buenos Aires.

Kobrin, Stephen J. 1987. "Testing the bargaining hypothesis in the manufacturing sector in developing countries." *International Organization* 41, no. 4 (Autumn): 609–38.

Krasner, Stephen D. 1976. "State Power and the Structure of International Trade." *World Politics* 28, no. 3 (Apr.): 317–47.

———. 1978. *Defending the National Interest: Raw Materials Investments and U.S. Foreign Policy*. Princeton: Princeton University Press.

———. 1984. "Approaches to the State: Alternative Conceptions and Historical Dynamics." *Comparative Politics* 16, no. 2:223–46.

Krauss, Ellis S. 1992. "Political Economy: Policymaking and Industrial Policy in Japan." *Political Science and Politics* 25 (Mar.): 44–56.

Kudrle, Robert T. 1985. "The Several Faces of the Multinational Corporation: Political Reaction and Policy Response." In W. Ladd Hollist and L. Tullis, eds., pp. 175–97.

Kuhn, James W. 1966. *Scientific and Managerial Manpower in the Nuclear Industry*. New York: Columbia University Press.

Kurth, James R. 1979. "The political consequences of the product cycle: industrial history and political outcomes." *International Organization* 33, no. 1 (Winter): 1–34.

Ladd, Everett C., Jr., and Seymour M. Lipset. 1972. "Politics of Academic Natural Scientists and Engineers." *Science* 176, June 9: 1091–1100.

Lakoff, Sanford A. 1977. "Scientists, Technologists, and Political Power." In Ina Spiegel-Rösing and D. De Solla Price, eds., *Science, Technology, and Society—A Cross-Disciplinary Perspective*, pp. 355–92. Beverly Hills, Calif.: Sage.

Lall, Sanjaya. 1980. *Developing Countries as Exporters of Technology: A First Look at the Indian Experience*. London: Macmillan.

———. 1991. "Multinational Enterprises and Developing Countries: Some Issues for Research in the 1990s." *Millennium Journal of International Studies* 20, no. 2 (Summer): 251–55.

Lamounier, Bolivar, and A. R. Moura. 1986. "Economic Policy and Political Opening in Brazil." In Jonathan Hartlyn and Samuel A. Morley, eds., *Latin American Political Economy*, pp. 168–92. Boulder, Colo.: Westview Press.

Landes, David S. 1992. "Homo Faber, Homo Sapiens: Knowledge, Technology, Growth, and Development." *Contention* 1, no. 3 (Spring): 81–108.

Langer, Erick D. 1989. "Generations of Scientists and Engineers: Origins of the Computer Industry in Brazil." *Latin American Research Review* 24, no. 2: 95–112.

Leff, Nathaniel H. 1968. *Economic Policy-Making and Development in Brazil*. New York: Wiley.

Leite, Rogerio C. de Cerqueira. 1978. *A agonia da tecnologia nacional*. São Paulo: Livraria Duas Cidades.

Lentner, Howard H. 1984. "The Concept of the State: A Response to Stephen Krasner." *Comparative Politics* 16, no. 3: 367–76.

Lepecki, W. P. S. 1985. "The Brazilian nuclear program and the technology transfer from the Federal Republic of Germany to Brazil: Nuclear power plant engineering and industrial promotion." Paper presented at the Second International Course on Nuclear Physics and Reactors, Bogotá, Colombia.

Levín, Emanuel. 1981. *Los jóvenes Argentinos y la investigación científico-tecnológica*. Buenos Aires: Ediciones Lihuel.

Levy-Leblond, Jean-Marc. 1976. "Ideology of/in Contemporary Physics." In H. Rose and S. Rose, eds., *The Radicalization of Science*, pp. 136–75. London: Macmillan.

Lewis, John W., H. Di, and Xue Litai. 1991. "Beijing's Defense Establishment: Solving the Arms-Export Enigma." *International Security* 15 (Spring): 87–109.

Lewis, John W., and Xue Litai. 1988. *China Builds the Bomb*. Stanford: Stanford University Press.

Lewis, Paul H. 1990. *The Crisis of Argentine Capitalism*. Chapel Hill: University of North Carolina Press.

Leys, Colin. 1977. "Underdevelopment and Dependency: Critical Notes." *Journal of Contemporary Asia* 7, no. 1:92–102.

———. 1984. "Relations of Production and Technology." In M. Fransman and K. King, eds., pp. 175–84.

Lichtenberg, Frank R. 1985. "Assessing the Impact of Federal Industrial R&D Expenditure on Private R&D Activity." Paper prepared for National Academies of Science and Engineering workshop "The Federal Role in R&D." Washington, D.C., Nov. 21–22.

Lima, Maria Regina Soares de. 1986. "The Political Economy of Brazilian Foreign Policy: Nuclear Energy, Trade, and Itaipú." Ph.D. diss., Vanderbilt University.

Lindblom, Charles E. 1968. *The Policy-Making Process*. Englewood Cliffs, N.J.: Prentice-Hall.

Lipset, Seymour M., and A. Solari. 1967. *Elites in Latin America*. New York: Oxford University Press.

Lopes, José Leite. 1972. *La ciencia y el dilema de América Latina: dependencia o liberación*. Buenos Aires: Siglo Veintuno.

———. 1978. *Ciencia e libertação*. Rio de Janeiro: Paz e Terra.

———. 1984. "Mario Schemberg: Lembranças en sua homenagem." Rio de Janeiro, Centro Brasileiro de Pesquisas Físicas, *Ciencia e Sociedade*, CBPF-cs-002/84.

Lovins, A., L. H. Lovins, and L. Ross. 1980. "Nuclear Power and Nuclear Bombs." *Foreign Affairs* (Summer): 1137–77.

Low, Morris F. 1994. "The Political Economy of Japanese Science: Nakasone, Physicists, and the State." In Etel Solingen, ed., pp. 93–120.

McDaniel, Tim. 1976–77. "Class and Dependency in Latin America." *Berkeley Journal of Sociology* 21:51–88.

McDonough, Peter. 1981. *Power and Ideology in Brazil*. Princeton: Princeton University Press.

Mallon, R. D., and J. V. Sourrouille. 1975. *Economic Policymaking in a Conflict Society: The Argentine Case*. Cambridge, Mass.: Harvard University Press.

March, James G., and Johan P. Olsen. 1989. *Rediscovering Institutions — The Organizational Basis of Politics*. New York: Free Press.

Mares, David. 1988. "Middle Powers Under Regional Hegemony: To Challenge or Acquiesce in Hegemonic Enforcement." *International Studies Quarterly* 32 (Dec.): 453–72.

Mariscotti, Mario. 1985. *El secreto atómico del Huemul*. Buenos Aires: Sudamericana/Planeta.

Martin, H. D., and J. Spitalnik. 1985. "Nuclebrás and KWU collaboration for simulator training." Paper presented at the Third International Conference on Nuclear Technology Transfer, Madrid, Oct.

Martínez Nogueira, R. 1976. "Las decisiones tecnológicas de las empresas públicas. Conclusiones de los estudios de caso." Serie Informes de Investigación 6, Buenos Aires: Economic Commission for Latin America.

Martins, Carlos E. 1977. *Capitalismo de estado e modelo político no Brasil*. Rio de Janeiro: Edições do Graal.

Martins, Luciano. 1986. "The 'Liberalization' of Authoritarian Rule in Brazil." In Guillermo O'Donnell, et al., pp. 72–94.

Marton, Katherine. 1986. *Multinationals, Technology and Industrialization*. Lexington, Mass.: Lexington Books.

Marx, Karl, and Friedrich Engels. 1968. *Selected Works*. Vol. 1. New York: International Publishers.

Mason, R. H. 1981. "Comments." In T. Sagafi-Nejad, et al., pp. 27–36.

Maxwell, P. M. 1981. "The Latin American experience." In R. S. Cohen, et al., *The social implications of the scientific and technological revolution*, pp. 230–51. Paris: UNESCO.

Meneses, L. C., and D. N. Simon. 1981. "Dois erros em cadeia: A política nuclear e a estrutura organizacional do programa nuclear brasileiro." In David N. Simon, et al., pp. 23–33.

Merton, Robert K. 1968. *Social Theory and Social Structure*. New York: Free Press.

———. 1973. *The Sociology of Science*. Chicago: University of Chicago Press.

Migdal, Joel S. 1988. *Strong Societies and Weak States*. Princeton: Princeton University Press.

Miliband, Ralph. 1983. *Class Power and State Power*. London: Verso.

Millán, Victor. 1986. "Argentina: Schemes for Glory." In Michael Brzoska and Thomas Ohlson, eds., *Arms Production in the Third World*. London: Taylor and Francis.

Milner, Helen. 1987. "Resisting the protectionist temptation: industry and the making of trade policy in France and the United States during the 1970s." *International Organization* 41, no. 4 (Autumn): 639–66.

———. 1988. *Resisting Protection: Global Industries and the Politics of International Trade*. Princeton: Princeton University Press.

Mirchandani, G. G., and P. K. S. Namboodiri. 1981. *Nuclear India: A Technological Assessment*. New Delhi: Vision Books.

Mirow, K. R. 1979. *Loucura nuclear* (os enganos do acordo Brasil-Alemanha). Rio de Janeiro: Civilização brasileira.

Mitra, J. D. 1979. *The Capital Goods Sector in LDCs: A Case Study for State Intervention*. Washington, D.C.: World Bank Staff Working Papers no. 343.

Montgomery, John D. 1981. "Development Without Tears—Can Science Policy Reverse the Historical Process?" In Joseph S. Szyliowicz, ed., pp. 151–72.

Moore, F. T. 1983. *Technological Change and Industrial Development: Issues and Opportunities*. Washington, D.C.: World Bank Working Paper no. 613.

Moran, Theodore H., ed. 1974. *Multinational corporations and the Politics of Dependence: Copper in Chile*. Princeton: Princeton University Press.

————, ed. 1978. "Multinational corporations and dependency: a dialogue for dependentistas and non-dependentistas." *International Organization* 32, no. 1 (Winter): 79–100.

————. 1985. *Multinational Corporations: The Political Economy of Foreign Direct Investment.* Lexington, Mass.: Lexington Books.

————, ed. 1986. *Investing in Development: New Roles for Private Capital?* New Brunswick, N.J.: Transaction Books.

Morel, R. L. de M. 1979. *Ciencia e estado: a política científica no Brasil.* São Paulo: T. A. Queiroz.

Morita-Lou, H., ed. 1985. *Science and Technology Indicators for Development.* Boulder, Colo.: Westview Press.

Most, Benjamin A. 1991. *Changing Authoritarian Rule and Public Policy in Argentina, 1930–1970.* Boulder, Colo.: Lynne Rienner.

Mounet, Jacques Richard. 1985. "Export financing in France." *IAEA Bulletin* 27, no. 4:41–43.

Mulkay, M. J. 1977. "Sociology of the Scientific Research Community." In I. Spiegel-Rösing and Derek de Solla Price, eds., pp. 93–148.

Muller, A. E. F., A. E. Gasparian, H. J. Calvet Filho. 1980. "Aspects of Consolidation of Engineering Capability Related to Nuclear Power Plants." Paper presented at the Conference on utilization of small- and medium-sized power reactors in Latin America. Montevideo, Uruguay, May 12–15.

Munck, Ronaldo. 1979. "State Intervention in Brazil: Issues and Debates." *Latin American Perspectives* 23, 6, 4.

————. 1984. *Politics and Dependency in the Third World: The Case of Latin America.* London: Zed Books.

Myers, David J. 1984. "Brazil: Reluctant Pursuit of the Nuclear Option." *Orbis* 27, no. 4 (Winter): 881–912.

Mytelka, Lynn K. 1979. *Regional Development in a Global Economy: The MNC, Technology, and Andean Integration.* New Haven: Yale University Press.

Nash, Nathaniel C. 1991. "Argentina shops shares in its telephone system." *New York Times*, Dec. 8.

————. 1992. "U.S. company wins stake in Argentine pipeline." *New York Times*, Dec. 4.

National Aeronautics and Space Administration (NASA). 1977–81. *NASA Spinoff.* Washington, D.C.: Technology Utilization Office.

Nau, Henry R. 1974. *National Politics and International Technology: Nuclear Reactor Development in Western Europe.* Baltimore: Johns Hopkins University Press.

————. 1990. *The Myth of America's Decline: Leading the World Economy into the 1990s.* New York: Oxford University Press.

Nelson, Richard R. 1974. "Less Developed Countries—Technology Transfer and Adaptation: The Role of the Indigenous Science Community." *Economic Development and Cultural Change* 23 (Oct.): 61–77.

Nelson, Richard R., and S. Winter. 1977. "In Search of a Useful Theory of Innovation." *Research Policy* 6, no. 1 (Jan.): 36–77.

Newfarmer, Richard S. 1978. "The international market power of transnational corporations: a case study of the electrical industry." Geneva: United Nations Commission on Trade and Development (UNCTAD).

———. 1985. "International Oligopoly in the Electrical Industry." In Richard S. Newfarmer, ed., *Profits, Progress, and Poverty: Case Studies of International Industries in Latin America*, pp. 113–50. Notre Dame, Ind.: University of Notre Dame Press.

Newfarmer, R. S., and W. F. Mueller. 1975. *Multinational Corporations in Brazil and Mexico: Structural Sources of Economic and Noneconomic Power.* Report to the Subcommittee on Multinational Corporations of the Committee on Foreign Relations. Washington, D.C.: U.S. Senate (Aug.).

Niskanen, William A., Jr. 1971. *Bureaucracy and Representative Government.* Chicago: Aldine-Atherton.

Nogués, Julio J. 1986. *The Nature of Argentina's Policy Reforms during 1976–81.* Washington, D.C.: World Bank, World Bank Staff Working Paper 765.

Nordlinger, Eric. 1981. *On the Autonomy of the Democratic State.* Cambridge, Mass.: Harvard University Press.

North, Douglass C. 1981. *Structure and Change in Economic History.* New York: Norton.

———. 1990. *Institutions, Institutional Change, and Economic Performance.* New York: Cambridge University Press.

Nuclebrás (Empresas Nucleares Brasileiras). 1980. *Relatório de atividades.* Rio de Janeiro.

———. 1981. *Relatório de atividades.* Rio de Janeiro.

———. 1982a. *Fábrica de elementos combustíveis (FEC).* Rio de Janeiro (Oct.).

———. 1982b. *Relatório de atividades.* Rio de Janeiro.

———. 1983. *Relatório de atividades.* Rio de Janeiro.

———. 1984. *Relatório de atividades.* Rio de Janeiro.

———. n.d. *O avanço da prospeção e pesquisa de urânio no Brasil.* Rio de Janeiro.

———. n.d. *O Brasil desenvolve a tecnologia nuclear.* Rio de Janeiro.

———. n.d. *O CIPC.* Rio de Janeiro.

———. n.d. *Pesquisa nuclear no CDTN — O dominio e fixação da tecnologia.* Rio de Janeiro.

Nussenzveig, H. M. 1969. "Migration of Scientists from Latin America." *Science* 165 (Sept.): 1328–32.

Nye, Joseph S. 1981. "Maintaining a Nonproliferation Regime." In George H. Quester, ed., *Nuclear Proliferation: Breaking the Chain*, pp. 15–38. Madison: University of Wisconsin Press.

———. 1990. *Bound to Lead: The Changing Nature of American Power.* New York: Basic Books.

Odell, John S. 1982. *United States International Monetary Policy—Markets, Power, and Ideas as Sources of Change*. Princeton: Princeton University Press.

O'Donnell, Guillermo. 1973. *Modernization and Bureaucratic-Authoritarianism: Studies in South American Politics*. Berkeley, Calif.: Institute of International Studies.

——. 1976. *Estado y alianzas en la Argentina: 1956–1976*. Buenos Aires, CEDES, G. E. CLACSO no. 5.

——. 1978. "Reflections on the Patterns of Change in the Bureaucratic-Authoritarian State." *Latin American Research Review* 13, no. 1:3–38.

——. 1988. *Bureaucratic Authoritarianism*. Berkeley: University of California Press.

O'Donnell, Guillermo, Philippe C. Schmitter, and L. Whitehead, eds. 1986. *Transitions from Authoritarian Rule: Latin America*. Baltimore: Johns Hopkins University Press.

Okimoto, Daniel I. 1989. *Between MITI and the Market: Japanese Industrial Policy for High Technology*. Stanford: Stanford University Press.

Oliveira, Eliézer Rizzo de. 1976. *As forças armadas: política e ideologia no Brasil (1964–1969)*. Petrópolis: Vozes.

Onis, Ziya. 1991. "The Logic of the Developmental State." *Comparative Politics* 24, no. 1 (Oct.): 109–26.

Oszlak, Oscar. 1976. "Política y organización estatal de las actividades científico-técnicas en la Argentina: crítica de modelos y prescripciones corrientes." Estudios Sociales no. 2. Buenos Aires: Centro de Estudios de Estado y Sociedad (CEDES).

——. 1984. "Políticas públicas y regímenes políticos." Buenos Aires: Centro de Estudios de Estado y Sociedad (CEDES).

——. 1986. "Public policies and political regimes in Latin America." *International Social Science Journal* 108:219–36.

Pack, Howard. 1979. "Technology and Employment: Constraints on Optimal Performance." In Samuel M. Rosenblatt, ed., pp. 59–86.

——. 1982. "The Capital Goods Sector in LDCs: Economic and Technical Development." In Moshe Syrquin and Simon Teitel, eds., pp. 349–72.

Packenham, Robert A. 1992. *The Dependency Movement—Scholarship and Politics in Development Studies*. Cambridge, Mass.: Harvard University Press.

——. 1994. "The Politics of Economic Liberalization: Argentina and Brazil in Comparative Perspective." Notre Dame, Ind.: University of Notre Dame: The Helen Kellogg Institute for International Studies, Working Paper no. 206 (Apr.).

Parker, J. E. S. 1978. *The Economics of Innovation—The National and Multinational Enterprise in Technological Change*. New York: Longman.

Payne, Leigh A. 1994. *Brazilian Industrialists and Democratic Change*. Baltimore, Md.: Johns Hopkins University Press.

Pempel, T. J. 1982. *Policy and Politics in Japan.* Philadelphia: Temple University Press.

Perlmutter, Howard V., and Tagi Sagafi-Nejad. 1981. *International Technology Transfer: Guidelines, Codes and a Muffled Quadrilogue.* New York: Pergamon Press.

Pescarmona, Luis. 1982. "Manufacturing of Primary Heavy Components: A New Experience in Argentina." *Transactions.* La Grange Park, Ill.: American Nuclear Society.

Petras, James. 1977. "State Capitalism and the Third World." *Development and Change* 8, 1.

Pinto, Ricardo Guedes F. n.d. *Liliputianos e Lapucianos: os caminhos da física no Brasil (1810 a 1949).* Grupo de Estudos sobre o Desenvolvimento da Ciencia, Centro de Estudos e Pesquisas.

Pion-Berlin, David. 1985. "The Fall of Military Rule in Argentina: 1976–1983." *Journal of Interamerican Studies and World Affairs* 27 (Summer): 55–76.

———. 1989. *The Ideology of State Terror: Economic Doctrine and Political Repression in Argentina and Peru.* Boulder, Colo.: Lynne Rienner.

Piore, Michael J., and Charles F. Sabel. 1984. *The Second Industrial Divide—Possibilities for Prosperity.* New York: Basic Books.

Pirro e Longo, W. 1984. *Tecnologia e soberania nacional.* São Paulo: Nobel, Promocet.

Polanyi, Michael. 1968. "The Republic of Science: Its Political and Economic Theory." In Edward Shils, ed., *Criteria for Scientific Development: Public Policy and National Goals,* pp. 1–20. Cambridge, Mass.: MIT Press.

Poneman, Daniel. 1982. *Nuclear Power in the Developing World.* Boston: Allen and Unwin.

———. 1984. "Nuclear Proliferation Prospects for Argentina." *Orbis* (Winter): 853–80.

———. 1987. *Argentina: Democracy on Trial.* New York: Paragon House.

Potash, Robert. 1980. *The Army and Politics in Argentina 1945–1962: Perón to Frondizi.* Stanford: Stanford University Press.

Potter, William C. 1982. *Nuclear Power and Nonproliferation.* Cambridge, Mass.: Oeleschlager, Gunn, and Hain.

Poulantzas, Nicos. 1975. *As classes sociais no capitalismo de hoje.* Rio de Janeiro: Zahar.

Poznanski, Kazimierz Z. 1984. "Technology Transfer: West-South Perspective." *World Politics* 37, no. 1 (Oct.): 134–52.

Price, Don K. 1965. *The Scientific Estate.* Cambridge, Mass.: Belknap Press.

Przeworski, Adam, and Henry Teune. 1970. *The Logic of Comparative Social Inquiry.* Malabar, Fla.: Krieger.

Putnam, Robert D. 1988. "Diplomacy and Domestic Politics: The Logic of Two-Level Games." *International Organization* 42 (Summer): 427–60.

Quester, George H. 1979. *Brazil and Latin-American Nuclear Proliferation: An Optimistic View.* Center for International and Strategic Affairs, University of California, Los Angeles, ACIS Working paper No. 17 (Dec.).

Quester, George H., ed. 1981. *Nuclear Proliferation*. Madison: University of Wisconsin Press.

Quijano, A. 1974. "Imperialism and International Relations in Latin America." In Julio Cotler and Richard Fagen, eds., *Latin America and the United States: The Changing Political Realities*. Stanford: Stanford University Press.

Ramamurti, Ravi. 1987. *State-Owned Enterprises in High Technology Industries: Studies in India and Brazil*. New York: Praeger.

Ramesh, J., and C. Weiss, eds. 1979. *Mobilizing Technology for World Development*. New York: Praeger.

Randall, Laura. 1978. *An Economic History of Argentina in the Twentieth Century*. New York: Columbia University Press.

Ranis, Gustav. 1984. "Determinants and Consequences of Indigenous Technological Activity." In M. Fransman and K. King, eds., pp. 95–112.

Rattner, Henrique, ed., 1979. *Brazil 1990 — caminos alternativos do desenvolvimento*. São Paulo: Brasiliense.

Redick, John R. 1978. "Regional Restraint: U.S. Nuclear Policy and Latin America." *Orbis* 22, no. 1: 161–200.

———. 1981. "The Tlatelolco Regime and Nonproliferation in Latin America." *International Organization* 35, no. 1 (Winter): 103–34.

———. 1988. *Nuclear Restraint in Latin America: Argentina and Brazil*. Programme for Promoting Nuclear Non-Proliferation. University of Southampton. Occasional Paper no. 1.

Relatório. 1984. *A Questão Nuclear Comissão Parlamentar de Inquérito do Senado Federal sobre o Acordo Nuclear do Brasil com a República Federal da Alemanha*. Vols. 1–6. Vol. 6 (Anexo: atas e depoimentos) is divided into nos. 1–6. Brasília: Senado Federal.

Remmer, Karen L. 1989. *Military Rule in Latin America*. Boston: Unwin Hyman.

Reppy, Judith. 1985. "Military R&D and the Civilian Economy." *Bulletin of the Atomic Scientists* (Oct.): 10–14.

Richardson, J. David. 1990. "The political economy of strategic trade policy." *International Organization* 44, no. 1 (Winter): 107–35.

Robinson, Austin, ed., 1979. *Appropriate Technologies for Third World Development*. Great Britain: International Economic Association.

Rock, David. 1987. "Political Movements in Argentina: A Sketch from Past and Present." In Mónica Peralta-Ramos and Carlos H. Waisman, eds., *From Military Rule to Liberal Democracy in Argentina*, pp. 3–20. Boulder, Colo.: Westview Press.

Rodrigo, F., O. Valentinuzzi, R. Costarelli, and A. E. Belluco. 1982. "Technology Transfer on Development and Production of Nuclear Raw Materials." Paper presented at the Second International Conference on Nuclear Technology Transfer, Buenos Aires.

Rosa, Luiz P., ed. 1978. *Energia, tecnologia e desenvolvimento: a questão nuclear*. Rio de Janeiro: Vozes.

———. 1984. *Energia e crise*. Petrópolis: Vozes.

———. 1985. "Que fazer da Nuclebrás." *Revista Brasileira de Tecnologia* 16, no. 2:61–62.

———. 1987. "Segurança dos reatores nucleares no Brasil." In José Carvalho, et al., eds., *O Brasil nuclear*, pp. 31–48. Porto Alegre: Tche!

Rosa, Luiz P., et al. 1984. "Technological and Economic Impacts of the Brazilian Nuclear Program." Rio de Janeiro: COPPE, UFRJ, Area Interdisciplinar de Energia. Mimeo.

Rosa, Luiz P., and J. B. S. Hesles. 1984. "Socio-Environmental Impacts of Brazil's Nuclear Program." Rio de Janeiro: COPPE, UFRJ (mimeo).

Rosecrance, Richard. 1990. *America's Economic Resurgence: A Bold New Strategy.* New York: Harper and Row.

Rosenbaum, H. Jon. 1975. "Brazil's Nuclear Aspirations." In Onkar Marwah and Ann Schulz, eds., *Nuclear Proliferation and the Near Nuclear Countries*, pp. 255–77. Cambridge, Mass.: Ballinger.

Rosenberg, Nathan. 1976. *Perspectives on Technology.* Cambridge, Eng.: Cambridge University Press.

———. 1982. *Inside the Black Box—Technology and Economics.* Cambridge, Eng.: Cambridge University Press.

———. 1986. "Civilian 'Spillovers' from Military R&D Spending: The American Experience Since World War II." Mimeo.

Rosenblatt, Samuel M., ed. 1979. *Technology and Economic Development: A Realistic Perspective.* Boulder, Colo.: Westview Press.

Rostow, W. W. 1960. *The Stages of Economic Growth: A Non-Communist Manifesto.* Cambridge, Eng.: Cambridge University Press.

Rouquié, Alain. 1981. *Poder militar y sociedad política en la Argentina.* Vol. 1, *Hasta 1943.* Buenos Aires: Emecé.

———. 1982. *Poder militar y sociedad política en la Argentina.* Vol. 2, *1943–1973.* Buenos Aires: Emecé.

———. 1987. *The Military and the State in Latin America.* Trans. Paul E. Sigmund. Berkeley: University of California Press.

Rowe, James W. 1969. "Science and Politics in Brazil: Background of the 1967 Debate on Nuclear Energy Policy." In Kalman H. Silvert, ed., *The Social Reality of Scientific Myth, Science and Social Change.* New York: American Universities Field Staff.

Rueschemeyer, D., and P. B. Evans. 1985. "The State and Economic Transformation: Toward an Analysis of the Conditions Underlying Effective Intervention." In Peter Evans, et al., eds., pp. 44–77.

Ruggie, John G. 1982. "International Regimes, Transactions, and Change: Embedded Liberalism in the Postwar Economic Order." *International Organization* 36 (Spring): 379–416.

———, ed. 1983. *The Antinomies of Interdependence.* New York: Columbia University Press.

Rycroft, Robert, and Joseph S. Szyliowicz. 1980. "The Technological Dimension of Decision-Making: The Case of the Aswan High Dam." *World Politics* 33, no. 1 (Oct.): 36–61.

Sábato, Jorge A. 1973. "Atomic Energy in Argentina: A Case-History." *World Development* 1, no. 8:23-39.

———. 1979. "Atomic Energy in Argentina: Toward Technological Autonomy." In J. Ramesh and C. Weiss, eds., pp. 144-53.

Sábato, Jorge A., and M. Mackenzie. 1982. *La producción de tecnología autónoma o transnacional.* Mexico D.F.: Nueva Imagen.

Sábato, Jorge A., O. Wortman, and G. Gargiulo. 1978. *Energía atómica e industria nacional.* Washington, D.C.: Organization of American States, Dept. of Scientific Affairs.

Sábato, Jorge Federico, and Jorge Schvarzer. 1983. "Funcionamiento de la economía y poder político en la Argentina: trabas para la democracia." Buenos Aires: Centro de Investigaciones Sociales sobre el Estado y la Administración.

Sagafi-Nejad, T., R. W. Moxon, and H. V. Perlmutter, eds. 1981. *Controlling International Technology Transfer.* New York: Pergamon Press.

Sagasti, Francisco R. 1978. *Science and Technology Policy Implementation in Less Developed Countries.* Ottawa: Project on Science and Technology Policy Implementation in Less Developed Countries.

———. 1979. *Technology, Planning, and Self-Reliant Development — A Latin American View.* New York: Praeger.

———. 1980. "Science, Technology and Development Planning: A Review of Key Issues." In K. H. Standke and M. Anandakrishnen, eds., *Science, Technology, and Society: Needs, Challenges and Limitations,* pp. 503-64. New York: Pergamon Press.

———. 1981a. *Ciencia, tecnología, y desarrollo latinoamericano.* Mexico: Fondo de Cultura Económica, Lecturas 42.

———. 1981b. *El factor tecnológico en la teoría del desarrollo económico.* Mexico: El Colegio de Mexico, Jornadas 94.

Sagasti, Francisco, and Alberto Aráoz. 1976. *Science and Technology Policy Implementation in Less-Developed Countries: Methodological Guidelines for the STPI Project.* Ottawa, Ontario: International Development Research Centre.

Samuels, Richard J. 1987. *The Business of the Japanese State — Energy Markets in Comparative and Historical Perspective.* Ithaca: Cornell University Press.

Sant'anna, Vanya M. 1978. *Ciencia e sociedade no Brasil.* São Paulo: Editora Símbolo.

Schatz, S. P. 1987. "Assertive Pragmatism and the Multinational Enterprise." In David G. Becker, et al., pp. 107-30.

Scheinman, Lawrence. 1965. *Atomic Energy Policy in France under the Fourth Republic.* Princeton: Princeton University Press.

———. 1972. "Security and a Transnational System: The Case of Nuclear Energy." In Robert O. Keohane and Joseph S. Nye, eds., *Transnational Relations and World Politics,* pp. 276-99. Cambridge, Mass.: Harvard University Press.

———. 1987. *The International Atomic Energy Agency and World Nuclear Order*. Washington, D.C.: Resources for the Future.

Schmitter, Philippe C. 1971. *Interest Conflict and Political Change in Brazil*. Stanford: Stanford University Press.

———. 1973. "The 'Portugalization' of Brazil?" In Alfred Stepan, *Authoritarian Brazil*, pp. 179–232.

Schneider, Ben Ross III. 1987. "Politics within the State: Elite bureaucrats and industrial policy in authoritarian Brazil." Ph.D. diss., University of California, Berkeley.

———. 1991. *Politics within the State: Elite bureaucrats and industrial policy in authoritarian Brazil*. Pittsburgh: University of Pittsburgh Press.

———. 1993. "The Career Connection: A Comparative Analysis of Bureaucratic Preferences and Insulation." *Comparative Politics* 25, no. 3 (Apr.): 331–50.

Schneider, Ronald M. 1976. *Brazil: Foreign Policy of a Future World Power*. Boulder, Colo.: Westview Press.

Schumpeter, Joseph A. 1961. *The Theory of Economic Development*. Cambridge, Mass.: Harvard University Press.

Schurr, Sam H., and Jacob Marschak. 1950. *Economic Aspects of Atomic Power*. Princeton: Princeton University Press.

Schvarzer, Jorge. 1983. "Martínez de Hoz: la lógica política de la política económica." Buenos Aires: Centro de Investigaciones Sociales sobre el Estado y la Administración (CISEA).

Schwartzman, Simon. 1978. *A formação da comunidade científica no Brasil*. São Paulo: FINEP, Ed. Nacional.

———. 1988. "High Technology Versus Self-Reliance: Brazil Enters the Computer Age." In Julian M. Chacel, P. S. Falk, and D. Fleischer, eds., *Brazil's Economic and Political Future*, pp. 67–82. Boulder, Colo.: Westview Press.

———. 1991. *A Space for Science: The Development of the Scientific Community in Brazil*. University Park: Penn. State University Press.

Sebastian, Luis de. 1979. "Appropriate Technology in Developing Countries: Some Political and Economic Considerations." In J. Ramesh and C. Weiss, eds., pp. 66–73.

Seers, Dudley. 1970. "The Stages of Economic Growth of a Primary Producer in the Middle of the 20th Century." In R. I. Rhodes, ed., *Imperialism and Underdevelopment*, pp. 163–80. New York: Monthly Review Press.

Sercovich, Francisco C. 1981. "The Exchange and Absorption of Technology in Brazilian Industry." In Thomas Bruneau and Phillipe Faucher, eds., pp. 127–40.

Serra, José. 1979. "Three Mistaken Theses Regarding the Connection between Industrialization and Authoritarian Regimes." In David Collier, ed., pp. 99–164.

Sheahan, John. 1987. *Patterns of Development in Latin America: Poverty, Repression, and Economic Strategy*. Princeton: Princeton University Press.

Shils, Edward. 1972. *The Intellectuals and the Powers and Other Essays.* Chicago: University of Chicago Press.

Sikkink, Kathryn. 1991. *Ideas and Institutions: Developmentalism in Brazil and Argentina.* Ithaca: Cornell University Press.

Silva, Darly H. da. 1984. *Development and Effectiveness of Scientific and Technological Research in Brazilian State Companies.* Brasília: Conselho Nacional de Pesquisas. Mimeo.

Simon, David N., et al. 1981. *Energia nuclear em questão.* Rio de Janeiro: Instituto Euvaldo Lodi, Coleção Universidade e Industria.

Simon, Herbert A. 1976. *Administrative Behavior.* New York: Free Press.

Skidmore, Thomas E. 1973. "Politics and Policy Making in Authoritarian Brazil, 1937–71." In Alfred Stepan, *Authoritarian Brazil*, pp. 3–46.

———. 1988. *The Politics of Military Rule in Brazil 1964–85.* New York: Oxford University Press.

Sklar, Richard. 1976. "Postimperialism: A Class Analysis of Multinational Corporate Expansion." *Comparative Politics* 9, no. 1 (Oct.): 75–92.

———. 1979. "The Nature of Class Domination in Africa." *Journal of Modern African Studies* 17, no. 4: 531–52.

Skocpol, Theda. 1979. *States and Social Revolutions.* Cambridge, Eng.: Cambridge University Press.

———. 1985. "Bringing the State Back In: Strategies of Analysis in Current Research." In Peter Evans, et al., eds., pp. 3–43.

Skolnikoff, Eugene B. 1993. *The Elusive Transformation: Science, Technology, and the Evolution of International Politics.* Princeton: Princeton University Press.

Smith, Roger K. 1987. "Explaining the nonproliferation regime: anomalies for contemporary international relations theory." *International Organization* 41, no. 2: 253–82.

Smith, William C. 1985. "Reflections on the political economy of authoritarian rule and capitalist reorganization in contemporary Argentina." In Philip O'Brien and Paul Cammack, eds., *Generals in Retreat — The Crisis of Military Rule in Latin America*, pp. 37–88. Manchester, U.K.: Manchester University Press.

———. 1991. *Authoritarianism and the Crisis of the Argentine Political Economy.* Stanford: Stanford University Press.

Smith, William C., Carlos H. Acuña, and Eduardo A. Gamarra, eds. 1994. *Latin American Political Economy in the Age of Neoliberal Reform: Theoretical and Comparative Perspectives for the 1990s.* New Brunswick, N.J.: Transaction Publishers.

Soares, Guido F. S. 1974. "Contribuição ao estudo da política nuclear brasileira." Ph.D. diss., Pontifícia Universidade Católica de São Paulo.

———. 1976. "O acordo de cooperação nuclear Brasil-Alemanha Federal." *Revista Forense* 253, year 72 (Jan.–Mar.): 207–32.

Sociedade Brasileira de Física. 1975. "A participação dos físicos no programa nuclear brasileiro." *Boletim* 6, Relatório da Comissão Especial da Sociedade Brasileira de Física. Rio de Janeiro.

Solingen, Etel. 1989. "Structures, Institutions, and Sectoral Adjustments: Arms and Nuclear Industries in Brazil and Argentina." Paper presented at the International Studies Association Annual Meeting, London, Apr.

———. 1990a. "Technology, Countertrade, and Nuclear Exports: Brazil as an Emerging Nuclear Supplier." In William C. Potter, ed., *International Nuclear Trade and Nonproliferation*, pp. 111–52. Lexington, Mass.: Lexington Books.

———. 1990b. "The Impact of Nuclear Programs on Industrial Technological Development." In Bernard Crousse, Jon Alexander, and Rejean Landry, eds., *Evaluating Science and Technology Policies*, pp. 129–54. Québec: Laval University Press.

———. 1991. "Managing Energy Vulnerability." *Comparative Strategy* 10, no. 2:177–99.

———. 1993. "Macropolitical Consensus and Lateral Autonomy in Industrial Policy: The Nuclear Sector in Brazil and Argentina." *International Organization* 47, no. 2 (Spring): 263–98.

———, ed. 1994a. *Scientists and the State: Domestic Structures and the International Context*. Ann Arbor: University of Michigan Press.

———. 1994b. "The Political Economy of Nuclear Restraint." *International Security* 19, no. 2 (Fall): 126–69.

Solinger, Dorothy J. 1991. *From Lathes to Looms: China's Industrial Policy in Comparative Perspective, 1979–1982*. Stanford, Calif.: Stanford University Press.

Solo, Robert. 1985. "The Formation and Transformation of States." In W. Ladd Hollist and Lamond Tullis, eds., pp. 69–86.

Sourrouille, Juan V. 1980. *Transnacionales en América Latina — el complejo automotor en Argentina*. Mexico D.F.: Editorial Nueva Imagen.

Souza, H. G., et al. 1972. *Política científica*. São Paulo: Editora Perspectiva.

Spero, Joan E. 1982. "Information: The Policy Void." *Foreign Policy* 48 (Fall): 139–56.

Spiegel-Rösing, I., and Derek de Solla Price, eds. 1977. *Science, Technology, and Society — A Cross-Disciplinary Perspective*. Beverly Hills, Calif.: Sage.

Spitalnik, Jorge. 1982. "Desenvolvimento de mão de obra para o programa nuclear brasileiro." Paper presented at the International Seminar on Energy Management, UNIDO, São Paulo, Oct.

Spitalnik, Jorge, and G. Fonseca. 1985. "Training in Brazil." Paper presented at the Fifth Pacific Basin Nuclear Conference, Seoul, Korea, May.

Spitalnik, Jorge, and C. Lerner. 1984. "Formação e treinamiento de pessoal para o programa nuclear da Nuclebrás." Paper presented at the Twelfth Scientific Conference, Buenos Aires, Dec.

Spitalnik, Jorge, C. Lerner, V. Stilben, and O. Botelho. 1984. "O programa de treinamento e desenvolvimento da Nuclebrás para o programa nuclear brasileiro." Paper presented at the Third Brazilian Congress on Energy, Rio de Janeiro, Oct.

Springer, Philip. 1968. "Disunity and Disorder: Factional Politics in the Argentine Military." In Henry Bienen, ed., *The Military Intervenes*, pp. 145–68. New York: Russell Sage Foundation.

Sprinzak, Ehud, and Larry Diamond, eds. 1993. *Israeli Democracy Under Stress*. Boulder, Colo.: Lynne Rienner.

Stepan, Alfred. 1971. *The Military in Politics: Changing Patterns in Brazil*. Princeton: Princeton University Press.

———, ed. 1973a. *Authoritarian Brazil: Origin, Policies, and Future*. New Haven: Yale University Press.

———. 1973b. "The New Professionalism of Internal Warfare and Military Role Expansion." In Alfred Stepan, ed., *Authoritarian Brazil*, pp. 47–68.

———. 1978. *The State and Society: Peru in Comparative Perspective*. Princeton: Princeton University Press.

———. 1985. "State Power and the Strength of Civil Society in the Southern Cone of Latin America." In Peter Evans, et al., eds., pp. 317–46.

———. 1988. *Rethinking Military Politics: Brazil and the Southern Cone*. Princeton: Princeton University Press.

———. 1989. *Democratizing Brazil — Problems of Transition and Consolidation*. New York: Oxford University Press.

Stepan, Nancy. 1976. *Beginnings of Brazilian Science*. New York: Neal Watson Academic Publishers.

Stevis, Dimitris, and Stephen P. Mumme. 1991. "Nuclear Power, Technological Autonomy, and the State in Mexico." *Latin American Research Review* 26, no. 3: 55–82.

Stewart, Frances. 1977. *Technology and Underdevelopment*. London: Macmillan.

———. 1981. "International Technology Transfer: Issues and Policy Options." In P. P. Streeten and R. Jolly, eds., pp. 67–110.

———. 1984. "Facilitating Indigenous Technical Change in Third World Countries." In M. Fransman and K. King, eds., pp. 81–94.

Stewart, Frances, and J. James, eds. 1982. *The Economics of New Technology in Developing Countries*. Boulder, Colo.: Westview Press.

Stiglitz, J. E. 1979. "On the Microeconomics of Technical Progress." Buenos Aires: Banco Interamericano de Desarrollo/CEPAL (Economic Commission for Latin America). Research Program Working Paper no. 32.

Stone, Jeremy J. 1977. "Brazilian Scientists and Students Resist Repression." *Public Interest Report* no. 30 (Nov.): 1–8. Federation of American Scientists.

———. 1984. "The Argentine Return to Democracy." *Journal of the Federation of American Scientists* (Apr.).

Strange, Susan. 1983. "Structures, Values, and Risk in the Study of the International Political Economy." In R. J. Barry Jones, ed., *Perspectives on Political Economy*, pp. 209–30. London: Frances Pinter.

———, ed. 1984. *Paths to International Political Economy*. London: George Allen & Unwin.

————. 1985. "International Political Economy: The Story So Far and the Way Ahead." In W. Ladd Hollist and Lamond Tullis, eds., pp. 13–26.

————. 1988. *States and Markets*. London: Pinter.

Street, James H. 1981. "Political Intervention and Science in Latin America." *Bulletin of Atomic Scientists* (Feb.): 16–23.

Streeten, P. P. 1979. *The Frontiers of Development Studies*. London: Macmillan.

Streeten, P. P., and R. Jolly, eds. 1981. *Recent Issues in World Development*. New York: Pergamon Press.

Sunkel, Osvaldo. 1971. "Capitalismo transnacional y desintegración nacional en América Latina." *Estudios Internacionales* 4, no. 16.

————. 1973. "The Pattern of Latin American Dependence." In Víctor L. Urquidi and Rosemary Thorp, eds., *Latin America in the International Economy*, pp. 3–25. London: Macmillan.

Surrey, John. 1987. "Electric power plant in India: A strategy of self-reliance." *Energy Policy* (Dec.): 503–21.

Surrey, John, and William Walker. 1981. *The European Power Plant Industry: Structural Responses to International Market Pressures*. Sussex European Research Center, University of Sussex.

Syllus, Carlos. 1982. "Brazil's Nuclear Policy and the Role of NUCLEN." *Nuclear Europe* 12:18–21.

Syrquin, Moshe, and Simon Teitel, eds. 1982. *Trade, Stability, Technology, and Equity in Latin America*. New York: Academic Press.

Szyliowicz, Joseph S., ed. 1981. *Technology and International Affairs*. New York: Praeger.

Tanis, Sara V. de. n.d. "Prospectiva de la demanda y desarrollo de productos metalúrgicos en la República Argentina." Buenos Aires: Comisión Nacional de Energía Atómica.

————. 1977. "Actividades de la Comisión Nacional de Energía Atómica." Buenos Aires: Comisión Nacional de Energía Atómica.

————. 1985. "Desarrollo de proveedores para la industria nuclear argentina." Buenos Aires: Comisión Nacional de Energía Atómica.

————. 1986. "Desarrollo del mercado nuclear argentino y de sus proveedores." Buenos Aires: Comisión Nacional de Energía Atómica.

Tanis, Sara V. de, and Y. S. de Andreone. 1984. "Transferencia de tecnología en la Comisión Nacional de Energía Atómica." Buenos Aires: Servicio de Asistencia a la Industria (SATI), Comisión Nacional de Energía Atómica.

Tanis, Sara V. de, and Jorge Kittl. 1976. "Veinte años de investigación y desarrollo." Buenos Aires: Comisión Nacional de Energía Atómica.

Tanis, Sara V. de, and M. R. Marrapodi. 1985. "Treinta años de patentamiento en el area nuclear." Buenos Aires: Comisión Nacional de Energía Atómica.

Tanis, Sara V. de, and Bennett Ramberg. 1990. "Argentina." In William C. Potter, ed., *International Nuclear Trade and Nonproliferation*, pp. 95–110. Lexington, Mass.: Lexington Books.

Tanis, Sara V. de, Jorge A. Sábato, and Zulema B. de Malik. 1980. "Desarrollo de Recursos Humanos en Metalurgia: Balance de una experiencia." Buenos Aires: Comisión Nacional de Energía Atómica, Programa Multinacional de Metalurgia.

Teitel, Simon. 1981. "Towards an Understanding of Technical Change in Semi-industrial Countries." *Research Policy* 10: 127–47.

———. 1984. "Technology Creation in Semi-industrial Economies." *Journal of Development Economics* 16: 39–61.

Tellez, Theresa. 1966. "The Crisis of Argentine Science." *Bulletin of Atomic Scientists* 22 (Dec.): 32–34.

Thome Filho, Zieli D. 1983. "Engenharia nuclear. Avaliação e perspectivas." Brasília: SEPLAN/Conselho Nacional de Pesquisas.

Timerbaev, Roland M., and Meggen M. Watt. 1995. *Inventory of International Nonproliferation Organizations and Regimes.* Monterey, Calif.: Monterey Institute of International Studies, Center for Nonproliferation Studies (Feb.).

Tollefson, Scott D. 1991. "Brazilian Arms Transfers, Ballistic Missiles, and Foreign Policy: The Search for Autonomy." Ph.D. diss. Baltimore, Md.: Johns Hopkins University Press.

Tooze, Roger. 1983. "'Sectoral Analysis' and the International Political Economy." In R. J. Barry Jones, ed., *Perspectives on Political Economy,* pp. 231–41. London: Frances Pinter.

———. 1984. "Perspectives and Theory: a Consumer's Guide." In Susan Strange, ed., pp. 1–22.

Trajtenberg, Manuel. 1983. "Dynamics and Welfare Analysis of Product Innovations." Ph.D. diss., Harvard University.

Trebat, T. J. 1981. "Public Enterprises in Brazil and Mexico: A Comparison of Origins and Performance." In Thomas Bruneau and Phillipe Faucher, eds., pp. 41–58.

———. 1983. *Brazil's State-Owned Enterprises: A Case-Study of the State as Entrepreneur.* Cambridge, Eng.: Cambridge University Press.

Trimberger, Ellen Kay. 1978. *Revolution from Above: Military Bureaucrats and Development in Japan, Turkey, Egypt, and Peru.* New Brunswick, N.J.: Transaction Books.

Tullock, Gordon. 1965. *The Politics of Bureaucracy.* Washington, D.C.: Public Affairs Press.

Turner, Frederick C. 1983. "Entrepreneurs and Estancieros in Perón's Argentina: Cohesion and Conflict Within the Elite." In Frederick C. Turner and José E. Miguens, eds., *Juan Perón and the Reshaping of Argentina,* pp. 223–36. Pittsburgh: University of Pittsburgh Press.

Tychojkij, J. 1982. "IRAM toward standardization in the nuclear field." Paper presented at the Second International Conference on Nuclear Technology Transfer, Buenos Aires.

Tyson, Laura D'Andrea. 1993. *Who's Bashing Whom? Trade Conflict in High Technology Industries.* Washington, D.C.: Institute for International Economics.

Vaitsos, Constantine V. 1979. "Government Policies for Bargaining with Transnational Enterprises in the Acquisition of Technology." In J. Ramesh and C. Weiss, eds., pp. 98–106.

Varas, Augusto. 1989. "Democratization and Military Reform in Argentina." In Augusto Varas, ed., *Democracy Under Siege: New Military Power in Latin America*, pp. 47–64. New York: Greenwood Press.

Vargas, José I. 1976. "Energia nuclear." *Ciencia e Cultura* 28, no. 9:1025–36.

Venancio, Filho F., et al. 1980. "Experience of a Brazilian architect-engineer in seismic analysis of nuclear structures components." Paper presented at a Conference on Utilization of Small- and Medium-sized Power Reactors in Latin America, Montevideo, Uruguay, May.

Vernon, Raymond. 1966. "International Investments and International Trade in the Product Cycle." *Quarterly Journal of Economics* 80, no. 1: 190–207.

———. 1971. *Sovereignty at Bay: The Multinational Spread of U.S. Enterprises*. New York: Basic Books.

———. 1977. *Storm Over the Multinationals: The Real Issues*. Cambridge, Mass.: Harvard University Press.

Vernon, Raymond, and Debora Spar. 1989. *Beyond Globalism: Remaking American Foreign Economic Policy*. New York: Free Press.

Vilas, Carlos M. 1972. "Dinámica del conflicto político y de la dominación social en la República Argentina." *Revista Latinoamericana de Ciencia Política* 3, no. 1:86–112.

Villela, Annibal Villanova. 1984. *Empresas do governo como instrumento de política econômica: os sistemas Siderbrás, Eletrobrás, Petrobrás, e Telebrás*. Rio de Janeiro: Instituto de Planejamento Econômico e Social/Instituto de Pesquisas.

Villela, Annibal V., and Werner Baer. 1980. *O setor privado nacional: problemas e políticas para seu fortalecimento*. Rio de Janeiro: Instituto de Planejamento Econômico e Social (IPEA)/Instituto de Pesquisas (INPES). Relatório de pesquisa no. 46.

Villela, Teotônio. 1976. "A façe oculta da estatização." *Visão* (June 28): 72–80.

Waisman, Carlos. 1987. *Reversal of Development in Argentina: Postwar Counterrevolutionary Policies and Their Structural Consequences*. Princeton: Princeton University Press.

Walker, W., and Mans Lonnroth. 1983. *Nuclear Power Struggles: Industrial Competition and Proliferation Control*. London: George Allen and Unwin.

Wallender, Harvey W. 1979. *Technology Transfer and Management in the Developing Countries*. Cambridge, Mass.: Ballinger Publishing Company.

Wasserman, A., et al. 1976. "A transferencia de tecnologia na industria petroquímica brasileira." Paper presented at the First Brazilian Congress on Petrochemicals, Rio de Janeiro, Nov.

Watson, Cynthia A. 1984. "Argentine Nuclear Development: Capabilities and Implications." Ph.D. diss., University of Notre Dame.

———. 1990. "Argentina." In Raju G. C. Thomas and Bennett Ramberg, eds., *Energy and Security in the Industrializing World*, pp. 101–22. Lexington: University Press of Kentucky.

Weir, Margaret. 1989. "Ideas and Politics: The Acceptance of Keynesianism in Britain and the United States." In Peter A. Hall, ed., pp. 53–86.

Welles, J. G., and R. H. Waterman, Jr. 1984. "Space Technology: Pay-off from Spin-off." *Harvard Business Review*, no. 4:106–18.

Wesson, Robert. 1981. *The United States and Brazil: Limits of Influence.* New York: Praeger Special Studies.

Westerkampf, José. 1982. "Scientific Freedom in Latin America: The Role of Governments." Paper presented at the Annual Meeting of the American Physical Society.

Westman, John. 1985. "Modern Dependency: A 'Crucial Case' Study of Brazilian Government Policy in the Minicomputer Industry." *Studies in Comparative International Development* 20, no. 6 (Summer): 25–47.

Weston, David. 1986. "The Economic Impact of Military R&D—A Literature Review." Unpublished ms.

Westphal, Larry E., L. Kim, and Carl J. Dahlman. 1985. "Reflections on the Republic of Korea's Acquisition of Technological Capability." In Nathan Rosenberg and Carlos Frischtak, eds., *International Technology Transfer—Concepts, Measures, and Comparisons*, pp. 167–221. New York: Praeger.

Westphal, Larry E., Y. W. Rhee, and G. Pursell. 1984. "Sources of Technological Capability in South Korea." In M. Fransman and K. King, eds., pp. 279–300.

Williams, Roger. 1984. "The International Political Economy of Technology." In Susan Strange, ed., pp. 70–90.

Wilsford, David. 1989. "Tactical Advantages versus Administrative Heterogeneity: The Strengths and the Limits of the French State." In James A. Caporaso, ed., pp. 128–72.

Wilson, James Q. 1989. *Bureaucracy: What Government Agencies Do and Why They Do It.* New York: Basic Books.

Wonder, Edward. 1977. "Nuclear Commerce and Nuclear Proliferation: Germany and Brazil, 1975." *Orbis* 21, no. 2:277–306.

World Bank Country Study. 1980. *Brazil: Industrial Policies and Manufactured Exports.* Washington, D.C.: World Bank.

Wynia, Gary W. 1978. *Argentina in the Postwar Era: Politics and Economic Policy Making in a Divided Society.* Albuquerque: University of New Mexico Press.

———. 1986. *Argentina: Illusions and Realities.* New York: Holmes and Meyer.

Ziman, J. M. 1982. "Social Responsibility of Scientists." *Interciencia* 7, no. 5:265–72.

Zysman, John. 1977. *Political Strategies for Industrial Order — State, Market, and Industry in France.* Berkeley: University of California Press.
————. 1983. *Governments, Markets, and Growth.* Ithaca: Cornell University Press.
Zysman, John, and Laura Tyson, eds. 1983. *American Industry in International Competition.* Ithaca: Cornell University Press.

Index

In this index an "f" after a number indicates a separate reference on the next page, and an "ff" indicates separate references on the next two pages. A continuous discussion over two or more pages is indicated by a span of page numbers, e.g., "57–59." "Passim" is used for a cluster of references in close but not consecutive sequence.

306   *Index*

Library of Congress Cataloging-in-Publication Data

Solingen, Etel, 1952–
   Industrial policy, technology, and international bargaining: designing nuclear
   industries in Argentina and Brazil / Etel Solingen
      p. cm.
      Includes bibliographical references and index.
   ISBN 0-8047-2601-9 (cloth)
      1. Nuclear industry—Government policy—Argentina.   2. Nuclear in-
   dustry—Government policy—Brazil.   3. Industrial policy—Argentina.
   4. Industrial policy—Brazil.   5. Technology transfer—Argentina.   6. Technology transfer—Brazil.   I. Title.
   HD9698.A72S65 1996
   333.792'4'0981—dc20
   95-23129   CIP

Original printing 1996
Last figure below indicates year of this printing:
05   04   03   02   01   00   99   98   97   96

Printed in the USA
CPSIA information can be obtained
at www.ICGtesting.com
JSHW021320221024
72173JS00001B/18

9 780804 726016